Ketamine

Kenji Hashimoto • Soichiro Ide • Kazutaka Ikeda
Editors

Ketamine

From Abused Drug to Rapid-Acting Antidepressant

Springer

Editors
Kenji Hashimoto
Division of Clinical Neuroscience
Chiba University
Chiba, Japan

Soichiro Ide
Tokyo Metropolitan Institute of Medical
Tokyo, Japan

Kazutaka Ikeda
Tokyo Metropolitan Institute of Medical
Tokyo, Japan

ISBN 978-981-15-2901-6 ISBN 978-981-15-2902-3 (eBook)
https://doi.org/10.1007/978-981-15-2902-3

This Springer imprint is published by the registered company Springer Nature Singapore Pte Ltd.
The registered company address is: 152 Beach Road, #21-01/04 Gateway East, Singapore 189721, Singapore

Contents

Ketamine Abuse: Past and Present

Ming-Chyi Huang and Shih-Ku Lin

Abstract Ketamine is an anesthetic derivative of phencyclidine with dissociative, analgesic, and psychedelic properties. It is extensively used as an anesthetic drug in surgical procedures in pediatric, obstetric, and geriatric patients, as well as in veterinary settings. Since its approval for clinical use, the misuse and abuse of ketamine have been reported in scientific journals and in popular media globally, particularly in the last two decades. It has been used as a common club drug in dance, rave, and squat party scenes. Owing to its unique pharmacological properties, ketamine has been used as an antidepressant for the treatment of resistant depression. However, a long-term use of high doses of ketamine leads to numerous physical and psychological negative effects, such as abdominal pain, urinary system disorder, dependence/tolerance/withdrawal, cognitive impairment, psychosis, and depression. At present, no effective pharmacotherapy for managing compulsive drug-seeking behavior in patients with ketamine use disorder is available. Treatment that incorporates regular urine screening in addition to medications for psychiatric symptoms and craving may present favorable results.

Keywords Abuse · K-hole · Physical consequence · Psychological consequence · Regulation · Treatment

Ketamine is an anesthetic derivative of phencyclidine (or *N*-1-phenylcyclohexylpiperidine [PCP]) with dissociative, analgesic, and psychedelic properties. In the 1950s, the staff at Parke-Davis Industries developed two cylcoheximide drugs, PCP and *N*-ethyl-1-phenylcyclohexylamine chlorhydrate, as ideal anesthetic agents with analgesic effects (Domino 2010). PCP was commercialized in the United States; however, it was withdrawn from medical use in 1978 because of its

M.-C. Huang · S.-K. Lin (✉)
Department of Addiction Sciences, Taipei City Psychiatric Center, Taipei City Hospital, Taipei, Taiwan

Department of Psychiatry, School of Medicine, College of Medicine, Taipei Medical University, Taipei, Taiwan
e-mail: sklin@tpech.gov.tw

K. Hashimoto et al. (eds.), *Ketamine*,
https://doi.org/10.1007/978-981-15-2902-3_1

"schizophrenomimetic" effects and abuse potential as a recreational drug (with a street name of "angel dust"). Due to the failure to synthesize and develop a less potent and rapidly metabolized derivative of PCP with lower "psychic effects," ketamine continues to be an alternative medicinal drug (Huang et al. 2014). Although classified as a hallucinogen in pharmacology, ketamine is extensively used in various anesthetic procedures in pediatric, obstetric, and geriatric patients due to its safety profile (White et al. 1982; Li and Vlisides 2016). In addition, the antidepressant effects of ketamine in the treatment of refractory depression have been reported in recent years (Berman et al. 2000; Krystal et al. 2013). In this chapter, we discuss ketamine abuse from past to present.

1 Ketamine Abuse

FDA approved ketamine as an anesthetic in 1970. In the following year, scholars raised the question that if ketamine is a hallucinogen, it may be abused or misused, similar to lysergic acid diethylamide (LSD) (Reier 1971). Although the nonmedical use of ketamine began on the West Coast of the United States in the early 1970s (Petersen and Stillman 1978), scientific reports of the phenomenon were very few until the early 1990s.

Ketamine was not considered a major drug of abuse in the past two decades, and relatively few cases were reported during the period (Hurt and Ritchie 1994; Jansen and Rracot-Cankovic 2001). Popular use of ketamine originated from the rave or party pill (ecstasy) – adulterant mixture (Jansen 1993). In Taiwan, the most common compounds found in the analyses of urine from people attending rave parties are 3,4-methylenedioxymethamphetamine (MDMA), methamphetamine, ephedrine, ketamine, acetaminophen, and caffeine (Lua et al. 2003). Following the ecstasy epidemic, ketamine use has become more popular in recent years globally, particularly as a party drug in dance, rave, and squat party scenes. It has also become popular in the UK (Morgan et al. 2004; Muetzelfeldt et al. 2008), Australia (Dillon et al. 2003), Hong Kong (Cheng et al. 2007; Ng et al. 2010), and Mainland China (Fang et al. 2006).

The Expert Committee on Drug Dependence of the WHO prereviewed ketamine in 2003 and conducted a critical review in 2006. The committee concluded that "this information was not sufficient to warrant scheduling." At a 2012 meeting, the committee decided that "bringing ketamine under international control is not appropriate." At the level of the European Union, in 2000, the European Commission concluded that it was not appropriate to introduce control measures and recommended further monitoring of ketamine use. In the latest World Drug Report by United Nations Office on Drugs Control 2019 (UNODC 2019), ketamine is classified under new psychoactive substances (NPSs), which are not under the control of international drug conventions but which may pose a threat to public health. The growing popularity of ketamine in South-East Asia could be, in part, due to its lower status in regulatory systems and lower price point as a substitute for the increasingly expensive "ecstasy" or methamphetamine.

In recent years, ketamine has also become commonly abused in Taiwan, particularly among the youth and adolescents (Chen et al. 2009; Lua et al. 2003; Yen et al. 2007). Studies have reported that most ketamine users in Taiwan concomitantly used ecstasy (MDMA). MDMA and ketamine detection rates have been as high as 76% and 47%, respectively, among rave party participants in Taiwan (Lua et al. 2003). In addition, a focus group discussion conducted with club drug users revealed that there was a special drug use sequence extensively adhered to by Taiwanese polydrug users (Leung et al. 2008). In a single drug use episode, MDMA was often the first drug used, followed by ketamine, and then marijuana. This unique sequence of polydrug use in a single episode is called "trinity." A long-term use of ketamine may lead to dependence, and this disorder is diagnosed as "Other Hallucinogen Use Disorder" according to DSM-5 and as "Hallucinogen Abuse" based on the ICD-10 Code F16.1.

2 Psychedelic Effects of Ketamine

Contrary to the clinical doses for anesthesia (2 mg/kg) and depression (0.5 mg/kg), the abuse dose is quite high (>3 mg/kg). Most abusers snort ketamine powder directly, and other routes include smoking, ingestion, and rarely injection (Carmona-Huerta et al. 2019). The relatively rapid onset of the effects on the brain (approximately 5 min) and the short half-life (approximately 1–2 h) are believed to enhance binging use and the abuse potential of ketamine.

Ketamine is a noncompetitive *N*-methyl-D-aspartate (NMDA) receptor antagonist. It impairs working, episodic, and semantic memory acutely and causes psychotogenic and dissociative effects in healthy volunteers receiving a single dose. At high doses, ketamine can cause cardiovascular and respiratory toxicity while producing psychotropic effects, which become the primary experience, particularly in recreational users. The effect demonstrates a linear dose–effect relationship ranging from referential ideation, dissociation, and depersonalization to psychotic experiences, including a sensation of feeling light, body distortion, absence of time sense, novel experiences of cosmic oneness, and out-of-body experiences, often referred to as the "K-hole" (Curran and Monaghan 2001). According to the results of a structured interview to examine initiation experiences and the positive and negative effects of ketamine on recreational users (Muetzelfeldt et al. 2008), the most appealing aspects of ketamine were "melting into the surrounding," "visual hallucinations," "out-of-body experiences," and "giggliness," whereas the unappealing effects were "memory loss" and "decreased sociability." A user becomes trapped in a state of detachment from their physical environment and loses senses of time, space, and balance, as well as verbal skills.

In a survey of 1614 ketamine users, the common withdrawal symptoms reported were fatigue, poor appetite, drowsiness, cravings, anxiety, dysphoria, tremors, palpitation, and sweating (Chen et al. 2014). In addition, female ketamine users self-reported significantly greater levels of severity in cognitive impairment and urinary discomfort than male users.

3 Physical Effects of Ketamine Abuse

Experimental results revealed that ketamine could induce neurotoxic events and promote neuronal apoptosis (Olney et al. 2002). Chronic ketamine users were reported to have reduced frontal gray matter (Liao et al. 2011) and white matter abnormalities (Liao et al. 2010).

Regular ketamine abusers have reported vague or intense abdominal pain, colloquially termed “K-cramps” (Jansen 2000). They often attempt to alleviate the pain by consuming more ketamine, which makes it even more difficult to quit ketamine use. Case studies have described the dilation of the common bile duct with a smooth tapered end, mimicking choledochal cysts (Wong et al. 2009). A recent study reported that 18 of 26 (69%) patients had fusiform dilatation of the common bile ducts without evidence of intrinsic or extrinsic obstruction, and the severity of bile duct dilatation was correlated with the duration of ketamine use (Yu et al. 2014). Although the exact mechanism of fusiform dilatation is unclear, it is potentially associated with NMDA receptor blockade in the smooth muscle. Ketamine may also act via the dorsal motor nucleus of the vagus nerve, which has fibers projecting into the gall bladder. Discontinuing ketamine use is key to managing the abdominal pain; otherwise, treatment options are nonspecific.

Another major physical side effect of long-term ketamine abuse is interstitial cystitis. This is a severe and chronic complication, and the symptoms include dysuria, increased frequency of small-volume micturition, suprapubic pain, and if severe, painful hematuria and obstructive nephropathy. The novel clinical phenomenon was first documented by Shahani et al. in 2007 when they reported a single case of a pediatric patient with ketamine-associated ulcerative cystitis (Shahani et al. 2007). Computed tomographic scans of the patient revealed marked thickening of the bladder wall, reduced bladder capacity, and perivesical inflammation, and cystoscopy revealed severe ulcerative cystitis. A successive study by Chu et al. found that 42 of 59 (71%) patients with a “ketamine bladder” had various degrees of epithelial inflammation, similar to that observed in the case of chronic interstitial cystitis (Chu et al. 2008).

Several hypotheses have been proposed to explain the cause of urological complications. First, ketamine and norketamine may have direct toxic effects on the bladder when excreted. Second, the effect of ketamine on the central nervous system may decrease the contractile responses of the smooth muscle of the bladder. Third, damage to the glycosaminoglycan layer of the urothelium and interstitial cells may cause inflammatory responses and induce interstitial fibrosis (Chu et al. 2008). The prevalence of cystitis and contracted bladder is believed to be high in individuals who use ketamine daily over an extended period. Approximately 30% of recreational ketamine users report having urinary tract symptoms when using ketamine (Muetzelfeldt et al. 2008), and nearly half of the frequent users seek medical attention for cystitis. The prognosis of the disease varies, and one-third of cases resolve after ketamine use is discontinued, one-third remain in the same state, and the remaining one-third worsen (Cottrell and Gillatt 2008). A special term, "walking-stick ureters," has been used to describe the segmental beading from ureteral strictures and the straightening of both ureters in a case report (Huang et al. 2011). Nevertheless, abstaining from ketamine use is the milestone of treatment, followed by the administration of mucosal protective agents, such as pentosan polysulfate, or hyaluronic acid. Surgical intervention, such as augmentation enterocystoplasty or cystectomy with conduit diversion, is generally considered the last resort in patients who have persistent symptoms and hematuria despite the aforementioned therapies and abstinence from ketamine. However, the treatment outcomes depend on the severity of disease progression, similar to the case of interstitial cystitis (Chen et al. 2011).

4 Psychological Consequences of Ketamine Abuse

Over the last two decades, many reports have explored the negative psychological effects of chronic ketamine abuse. Most studies have focused on cognitive impairment; some have highlighted that chronic ketamine abuse may result in dependence, psychosis, and depression. In the following sections, we address the psychological consequences associated with chronic and heavy ketamine use.

4.1 Dependence

The peak effect of ketamine occurs within 20 min, with the desired effects dissipating within 90 min; the distribution half-life is 7–11 min, and the plasma half-life is approximately 2–4 h (Dillon et al. 2003; Moore and Bostwick 1999). The short half-life and pharmacokinetic properties of ketamine make it attractive but may also lay the ground for its abuse potential. The problem of ketamine dependence has gradually gained attention since the 1980s and has been highlighted in many case reports (Hurt and Ritchie 1994; Jansen 1990; Moore and Bostwick 1999; Pal et al. 2002; Kamaya and Krishna 1987; Ahmed and Petchkovsky 1980).

Animal studies have demonstrated that ketamine reinforces self-administration behavior and conditioned place preference (de Luca and Badiani 2011; Venniro et al. 2015). Acute treatment with ketamine increases extracellular dopamine concentrations, a key neurotransmitter underlying the reinforcing effects of drug abuse in the neural network of the reward system (the striatum and prefrontal cortex) (Breier et al. 1997; Smith et al. 1998; Vollenweider et al. 1997). It has been hypothesized that ketamine blocks NMDA glutamate receptors on gamma-aminobutyric acid (GABA) neurons, which results in the disinhibition of dopaminergic neurons and the subsequent increase in dopamine release (Liu et al. 2016). Notably, the results of a recent study indicated that even at low doses, repeated administration of ketamine could enhance addiction-like behavior (Strong et al. 2017).

In humans, data from imaging studies have revealed that ketamine could cause the striatal release of dopamine. The reinforcing and rewarding properties may explain the propensity for ketamine addiction and the increasing ketamine abuse trend globally over the past few decades (Li et al. 2011; Bokor and Anderson 2014). In addition, stress-related systems, such as the hypothalamus–pituitary–adrenal gland axis, and orexin (Huang et al. 2019) and oxytocin (Huang et al. 2018) signaling pathways, which have been postulated to play a role in the neuroadaptational mechanisms underlying addiction development, have been reported to be dysregulated in ketamine-dependent patients.

4.2 Tolerance

On the basis of serial investigations on ketamine abuse, Morgan et al. first highlighted that tolerance is a common feature in individuals who used ketamine frequently, and tolerance tend to escalate the dose over time. They reported that there might be a 600% increase in the ketamine dose compared with their initial dose of the drug (Morgan et al. 2008). They proposed that tolerance may be due to the induction of liver enzymes that metabolize ketamine (Livingston and Waterman 1978) as well as neuroadaptational changes that occur following chronic ketamine exposure (Morgan et al. 2012). Clinical case reports have also revealed that ketamine abusers develop considerable tolerance to the drug with very high doses (reaching 2.5 g/day) (Goyal et al. 2014; Critchlow 2006; Moore and Bostwick 1999; Pal et al. 2002). In treatment-seeking ketamine-dependent patients, ketamine use at doses as high as 4 g per day have been observed before the seeking treatment (Huang et al. 2018, 2019; Cheng et al. 2018; Wang et al. 2018).

4.3 Withdrawal

No consensus has been reached yet regarding whether a specific withdrawal syndrome exists following the discontinuation of chronic and heavy ketamine use. In general, physical withdrawal symptoms are not considered to be prominent in case reports describing ketamine dependence (Goyal et al. 2014); however, some reports have highlighted that somatic and psychological symptoms such as anxiety, depression, shaking, sweating, palpitations, or sleep impairment may be substantial in some users (Critchlow 2006; Morgan et al. 2008; Lin et al. 2016). Future studies that systemically examine the symptoms and signs in individuals with ketamine dependence should elucidate the clinical profile and characteristics of ketamine withdrawal.

5 Cognitive Impairment

Cognitive impairment is among the multiple adverse outcomes associated with ketamine abuse (Bokor and Anderson 2014; Morgan et al. 2010, 2012). The pharmacological effects of ketamine via a blockade of NMDA glutamate receptors, which is key to the mechanisms underlying learning and memory, could account for the disruption of cognitive function (Morgan et al. 2012). In addition, the effects on other neurotransmitter systems, for example, the inhibition of muscarinic acetylcholine receptor function (Durieux 1995) and induction of dopamine release (Rabiner 2007), could play a role.

A longitudinal study observed that increasing ketamine use over a year was correlated with decreasing performance in cognitive tasks, mainly spatial working memory and pattern recognition memory, suggesting that continued sustained ketamine use is harmful to cognitive function (Morgan et al. 2010). In addition, individuals who used ketamine actively (i.e., using ketamine 1–4 times per week and having the last dose within the preceding 1 month) exhibited deficits over an array of cognitive tests involving mental and motor speed, visual and verbal memory, and executive functions (Morgan et al. 2010; Tang et al. 2013), and the deficits seemed persistent (Morgan et al. 2010).

The cognitive function of treatment-seeking ketamine-dependent patients (i.e., those with a higher severity of ketamine use) was worse substantially compared with that of controls based on many cognitive tasks, including verbal memory, motor speed, verbal fluency, attention, and processing speed, as well as measures from the cognitive battery as a whole (Wang et al. 2018). Data from imaging studies have indicated that ketamine-dependent patients displayed damage in multiple brain areas, such as the frontal, parietal, occipital, limbic, and corpus striatum (Liao et al. 2011; Wang et al. 2013). In addition, the damage appears to be correlated dose dependently with the duration and cumulative dose of ketamine used (Liao et al. 2011), manifesting minute patches in the first year, and the patches became larger sites of atrophy by 4 years of heavy use (Wang et al. 2013). One animal study also showed that brain function was not altered after 1 month of ketamine use but was impaired significantly after 6 months of use (Sun et al. 2014). The observation of a correlation between the dose and cognitive deficit is in line with clinical reports that higher doses of ketamine use are associated with poorer verbal fluency (Wang et al. 2018) and that heavier lifetime ketamine use is correlated with verbal learning and verbal memory deficits (Chan et al. 2013). Such observations collectively suggest that chronic and heavy ketamine use might predispose individuals to more severe cognitive impairment, particularly under higher doses and longer administration. Whether the cognitive impairment would ameliorate following abstinence remains unknown despite evidence indicating a trend where a longer duration of abstinence is associated with better cognitive performance (Wang et al. 2018). In addition, the cognitive impairment associated with current ketamine users was not observed in ex-users (Tang et al. 2013). Future investigations that track the alterations in cognitive ability following more extended periods of abstinence from ketamine use are required.

6 Psychosis

Previous evidence demonstrating similarity between the cognitive and behavioral effects of ketamine in animals and humans and the signs and symptoms of schizophrenia (SZ) in humans suggests that abnormalities in NMDA receptor function play a role in the etiology of SZ (Javitt et al. 2012; Abi-Saab et al. 1998; Krystal et al. 2003; Lahti et al. 2001). Recreational ketamine abuse is associated with some subtle psychopathology that resolves with abstinence (Morgan et al. 2012). Persistent mild delusional ideation (Morgan et al. 2009, 2010), sensory disturbances, and other subthreshold psychotic symptoms (Fine and Finestone 1973; Stone et al. 2014; Tang et al. 2015a) are common among individuals with chronic ketamine abuse. Approximately 3% of ketamine abusers even develop psychotic symptoms that persist far beyond the period of intoxication (i.e., beyond 2 h following drug administration) (Krystal et al. 1994; Kleinloog et al. 2015; Abi-Saab et al. 1998; Adler et al. 1998; Liang et al. 2015; Zhang et al. 2014). The symptom factor structure was similar to that of SZ to a greater extent in individuals with ketamine-induced psychosis than in healthy individuals receiving a single dose of ketamine (Xu et al. 2015).

In a recent study, chronic heavy ketamine abusers with persistent psychotic symptoms beyond ketamine discontinuation (KPP) had PANSS profile total scores or subscale (positive, negative, and general psychopathology) scores similar to those of SZ patients (Cheng et al. 2018). However, those without persistent psychotic symptoms (KNP) had significantly less severe symptoms compared with the patients with KPP or SZ. In addition, individuals with KNP performed better in cognitive tests than the patients with KPP or SZ, who exhibited similar cognitive impairment characteristics, in particular spatial problem solving and verbal memory. Spatial memory has been found to be impaired in SZ patients, and this impairment predicts the transition to overt psychosis in individuals with a high risk of psychosis (Bang et al. 2015). In addition, patients with KPP consumed less than two-thirds of the daily ketamine consumption of patients with KNP. Overall, such observations suggest that ketamine administration might unmask a latent vulnerability to SZ-like syndrome in a subset of ketamine users and support the glutamatergic dysregulation hypothesis of SZ. Whether vulnerability to KPP, for example, based on genetic factors, might be a manifestation of risk factors unique to the syndrome or shared with SZ should be clarified in future studies.

7 Depression

Recent research has revealed the rapid antidepressant effects of acute intravenous infusion of low-dosed ketamine (Berman et al. 2000; Murrough et al. 2013; Zarate et al. 2006). The effects are superior with a repeated treatment protocol as compared with a single infusion (Murrough et al. 2013). In addition, the antidepressant effects generated by low anesthetic doses (10 mg/kg) of ketamine were not observed in rats receiving high doses (80 mg/kg) (Trujillo et al. 2011). In addition, preclinical evidence indicated that acute ketamine exposure would produce different neurobiological alterations following prolonged exposure (Kittelberger et al. 2012), suggesting that long-term and heavy ketamine abuse might be associated with distinct features resulting from the acute effects of low-dosed ketamine.

Some studies have suggested a link between chronic ketamine use and depression (Kalsi et al. 2011; Morgan et al. 2010, 2012). In a 1-year follow-up study, individuals who used ketamine frequently were observed to have increased depression scores over 12 months (Morgan et al. 2010). In clinical samples of ketamine abusers, self-reported depressive symptoms are rather common, with as high as 72.5%–77.5% of them scoring above the cut-off scores of moderate to severe depression assessed based on the Beck Depression Inventory (Fan et al. 2016; Tang et al. 2013). The prevalence of major depressive disorder (MDD) in outpatient samples of ketamine abusers was 7.8% (Liang et al. 2015), and the prevalence was estimated to be 18.5% through a chart-review analysis (Tang et al. 2015b), based on a standardized instrument. For inpatient populations, our data, also based a standardized interview, revealed that 23.3% (nearly one-fourth) of treatment-seeking ketamine-dependent patients had the MDD comorbidity (Chen et al. 2020).

To date, compared with its antidepressant effects, considerably less evidence is available with regard to the pro-depressant effects of ketamine (Chang et al. 2016). Contradictory outcomes for long-term to acute effects have been observed in other abused substances. For instance, although cocaine acutely produces intense euphoria, chronic administration of the drug may gradually cause a panic level of anxiety or depression (Rounsaville 2004). Because comorbid depression plays a role in conferring a risk of a poorer treatment response, more adverse outcomes (Hedden et al. 2010), maintaining substance use behavior (Cheetham et al. 2010; Cohn et al. 2014), assessment and screening of depressive symptoms, should be highlighted in the clinical management of chronic ketamine abusers.

8 Treatment of Ketamine Abuse

Currently, there is no effective pharmacotherapy for managing compulsive drug-seeking behavior in patients with ketamine use disorder. Ketamine abusers usually do not seek treatment until they have developed psychological problems or physical disorders, or they are forced to visit a psychiatric clinic for cessation use under the pressure of their family or law enforcement. In the search for substance use disorder treatment options, agents with "anticraving" properties have been proposed as an emerging class of psychotropic medication (O'Brien 2005). The antiepileptic lamotrigine has been proposed for the treatment of ketamine use disorder considering its pharmacological effect on glutamate release inhibition and the results of clinical case studies (Huang et al. 2016). Treatment that incorporates regular urine screening in addition to medication for psychiatric symptoms has revealed fair results. As ketamine abuse and dependence grow, addiction services need to be better informed on the drug and its effects. Additional studies should aim to address the health challenges experienced by this group of abusers and to explore best practices for the treatment of ketamine addiction.

9 Conclusion

Ketamine has had a turbulent history since its first use as a clinical anesthesia for humans in 1964 (Domino 2010). Considering its analgesic–anesthetic mechanisms and antidepressant effects, ketamine is associated with a high potential of abuse or dependence. The pros and cons of ketamine use as well as other uncertainties regarding its use (e.g., antidepressant effect) required further study in the future.

Ketamine users seldom seek help despite the awareness about the negative side effects of the drug, which may contribute to the severity of ketamine dependence that is complicated by the associated physical and mental harm. Therefore, additional studies are required to increase the availability and accessibility of effective treatment interventions as well as preventative public education.

References

Abi-Saab WM, D'Souza DC, Moghaddam B, Krystal JH (1998) The NMDA antagonist model for schizophrenia: promise and pitfalls. Pharmacopsychiatry 31(Suppl 2):104–109

Adler CM, Goldberg TE, Malhotra AK, Pickar D, Breier A (1998) Effects of ketamine on thought disorder, working memory, and semantic memory in healthy volunteers. Biol Psychiatry 43:811–816

Ahmed SN, Petchkovsky L (1980) Abuse of ketamine. Br J Psychiatry 137:303

Bang M, Kim KR, Song YY, Baek S, Lee E, An SK (2015) Neurocognitive impairments in individuals at ultra-high risk for psychosis: who will really convert? Aust N Z J Psychiatry 49:462–470

Berman RM, Cappiello A, Anand A, Oren DA, Heninger GR, Charney DS, Krystal JH (2000) Antidepressant effects of ketamine in depressed patients. Biol Psychiatry 47:351–354

Bokor G, Anderson PD (2014) Ketamine: an update on its abuse. J Pharm Pract 27:582–586

Breier A, Malhotra AK, Pinals DA, Weisenfeld NI, Pickar D (1997) Association of ketamine-induced psychosis with focal activation of the prefrontal cortex in healthy volunteers. Am J Psychiatry 154:805–811

Carmona-Huerta J, Castiello-De Obeso S, Ramirez-Palomino J, Duran-Gutierrez R, Cardona-Muller D, Grover-Paez F, Fernandez-Dorantes P, Medina-Davalos R (2019) Polypharmacy

in a hospitalized psychiatric population: risk estimation and damage quantification. BMC Psychiatry 19:78
Chan KW, Lee TM, Siu AM, Wong DP, Kam CM, Tsang SK, Chan CC (2013) Effects of chronic ketamine use on frontal and medial temporal cognition. Addict Behav 38:2128–2132
Chang H, Huang MC, Chen LY (2016) Major depressive disorder induced by chronic ketamine abuse: a case report. Prim Care Companion CNS Disord:18
Cheetham A, Allen NB, Yucel M, Lubman DI (2010) The role of affective dysregulation in drug addiction. Clin Psychol Rev 30:621–634
Chen LY, Chen CK, Chen CH, Chang HM, Huang MC, Xu K (2020) Association of craving and depressive symptoms in ketamine-dependent patients undergoing withdrawal treatment. Am J Addict 29(1):43–50
Chen WJ, Fu TC, Ting TT, Huang WL, Tang GM, Hsiao CK, Chen CY (2009) Use of ecstasy and other psychoactive substances among school-attending adolescents in Taiwan: national surveys 2004–2006. BMC Public Health 9:27
Chen CH, Lee MH, Chen YC, Lin MF (2011) Ketamine-snorting associated cystitis. J Formos Med Assoc 110:787–791
Chen WY, Huang MC, Lin SK (2014) Gender differences in subjective discontinuation symptoms associated with ketamine use. Subst Abuse Treat Prev Policy 9:39
Cheng WC, Ng KM, Chan KK, Mok VK, Cheung BK (2007) Roadside detection of impairment under the influence of ketamine—evaluation of ketamine impairment symptoms with reference to its concentration in oral fluid and urine. Forensic Sci Int 170:51–58
Cheng WJ, Chen CH, Chen CK, Huang MC, Pietrzak RH, Krystal JH, Xu K (2018) Similar psychotic and cognitive profile between ketamine dependence with persistent psychosis and schizophrenia. Schizophr Res 199:313–318
Chu PS, Ma WK, Wong SC, Chu RW, Cheng CH, Wong S, Tse JM, Lau FL, Yiu MK, Man CW (2008) The destruction of the lower urinary tract by ketamine abuse: a new syndrome? BJU Int 102:1616–1622
Cohn AM, Cobb C, Hagman BT, Cameron A, Ehlke S, Mitchell JN (2014) Implicit alcohol cognitions in risky drinking nicotine users with and without co-morbid major depressive disorder. Addict Behav 39:797–802
Cottrell AM, Gillatt D (2008) Consider ketamine misuse in patients with urinary symptoms. Practitioner 252:5
Critchlow DG (2006) A case of ketamine dependence with discontinuation symptoms. Addiction 101:1212–1213
Curran HV, Monaghan L (2001) In and out of the K-hole: a comparison of the acute and residual effects of ketamine in frequent and infrequent ketamine users. Addiction 96:749–760
de Luca MT, Badiani A (2011) Ketamine self-administration in the rat: evidence for a critical role of setting. Psychopharmacology (Berl) 214:549–556
Dillon P, Copeland J, Jansen K (2003) Patterns of use and harms associated with non-medical ketamine use. Drug Alcohol Depend 69:23–28
Domino EF (2010) Taming the ketamine tiger. Anesthesiology 113:678–684
Durieux ME (1995) Inhibition by ketamine of muscarinic acetylcholine receptor function. Anesth Analg 81:57–62
Fan N, Xu K, Ning Y, Rosenheck R, Wang D, Ke X, Ding Y, Sun B, Zhou C, Deng X, Tang W, He H (2016) Profiling the psychotic, depressive and anxiety symptoms in chronic ketamine users. Psychiatry Res 237:311–315
Fang YX, Wang YB, Shi J, Liu ZM, Lu L (2006) Recent trends in drug abuse in China. Acta Pharmacol Sin 27:140–144
Fine J, Finestone SC (1973) Sensory disturbances following ketamine anesthesia: recurrent hallucinations. Anesth Analg 52:428–430
Goyal S, Ambekar A, Ray R (2014) Ketamine dependence in an anesthesiologist: an occupational hazard? Indian J Psychol Med 36:335–337

Hedden SL, Martins SS, Malcolm RJ, Floyd L, Cavanaugh CE, Latimer WW (2010) Patterns of illegal drug use among an adult alcohol dependent population: results from the National Survey on Drug Use and Health. Drug Alcohol Depend 106:119–125
Huang PW, Meng E, Cha TL, Sun GH, Yu DS, Chang SY (2011) Walking-stick ureters' in ketamine abuse. Kidney Int 80:895
Huang MC, Lin SK, Chen CK (2014) A roller coaster from methamphetamine to ketamine. Taiwan J Psychiatry (Taipei) 28:217–231
Huang MC, Chen LY, Chen CK, Lin SK (2016) Potential benefit of lamotrigine in managing ketamine use disorder. Med Hypotheses 87:97–100
Huang MC, Chen LY, Chang HM, Liang XY, Chen CK, Cheng WJ, Xu K (2018) Decreased blood levels of oxytocin in ketamine-dependent patients during early abstinence. Front Psych 9:633
Huang MC, Chen CH, Chen LY, Chang HM, Chen CK, Lin SK, Xu K (2019) Chronic ketamine abuse is associated with orexin-A reduction and ACTH elevation. Psychopharmacology (Berl)
Hurt PH, Ritchie EC (1994) A case of ketamine dependence. Am J Psychiatry 151:779
Jansen KL (1990) Ketamine—can chronic use impair memory? Int J Addict 25:133–139
Jansen KL (1993) Non-medical use of ketamine. BMJ 306:601–602
Jansen KL (2000) A review of the nonmedical use of ketamine: use, users and consequences. J Psychoactive Drugs 32:419–433
Jansen KL, Rracot-Cankovic R (2001) The nonmedical use of ketamine, part two: a review of problem use and dependence. J Psychoactive Drugs 33:151–158
Javitt DC, Zukin SR, Heresco-Levy U, Umbricht D (2012) Has an angel shown the way? Etiological and therapeutic implications of the PCP/NMDA model of schizophrenia. Schizophr Bull 38:958–966
Kalsi SS, Wood DM, Dargan PI (2011) The epidemiology and patterns of acute and chronic toxicity associated with recreational ketamine use. Emerg Health Threats J 4:7107
Kamaya H, Krishna PR (1987) Ketamine addiction. Anesthesiology 67:861–862
Kittelberger K, Hur EE, Sazegar S, Keshavan V, Kocsis B (2012) Comparison of the effects of acute and chronic administration of ketamine on hippocampal oscillations: relevance for the NMDA receptor hypofunction model of schizophrenia. Brain Struct Funct 217:395–409
Kleinloog D, Uit Den Boogaard A, Dahan A, Mooren R, Klaassen E, Stevens J, Freijer J, Van Gerven J (2015) Optimizing the glutamatergic challenge model for psychosis, using S+ − ketamine to induce psychomimetic symptoms in healthy volunteers. J Psychopharmacol 29:401–413
Krystal JH, Karper LP, Seibyl JP, Freeman GK, Delaney R, Bremner JD, Heninger GR, Bowers MB Jr, Charney DS (1994) Subanesthetic effects of the noncompetitive NMDA antagonist, ketamine, in humans. Psychotomimetic, perceptual, cognitive, and neuroendocrine responses. Arch Gen Psychiatry 51:199–214
Krystal JH, D'Souza DC, Mathalon D, Perry E, Belger A, Hoffman R (2003) NMDA receptor antagonist effects, cortical glutamatergic function, and schizophrenia: toward a paradigm shift in medication development. Psychopharmacology (Berl) 169:215–233
Krystal JH, Sanacora G, Duman RS (2013) Rapid-acting glutamatergic antidepressants: the path to ketamine and beyond. Biol Psychiatry 73:1133–1141
Lahti AC, Weiler MA, Tamara Michaelidis BA, Parwani A, Tamminga CA (2001) Effects of ketamine in normal and schizophrenic volunteers. Neuropsychopharmacology 25:455–467
Leung KS, Li JH, Tsay WI, Callahan C, Liu SF, Hsu J, Hoffer L, Cottler LB (2008) Dinosaur girls, candy girls, and trinity: voices of Taiwanese club drug users. J Ethn Subst Abuse 7:237–257
Li L, Vlisides PE (2016) Ketamine: 50 years of modulating the mind. Front Hum Neurosci 10:612
Li JH, Vicknasingam B, Cheung YW, Zhou W, Nurhidayat AW, Jarlais DC, Schottenfeld R (2011) To use or not to use: an update on licit and illicit ketamine use. Subst Abuse Rehabil 2:11–20
Liang HJ, Tang KL, Chan F, Ungvari GS, Tang WK (2015) Ketamine users have high rates of psychosis and/or depression. J Addict Nurs 26:8–13

Liao Y, Tang J, Ma M, Wu Z, Yang M, Wang X, Liu T, Chen X, Fletcher PC, Hao W (2010) Frontal white matter abnormalities following chronic ketamine use: a diffusion tensor imaging study. Brain 133:2115–2122

Liao Y, Tang J, Corlett PR, Wang X, Yang M, Chen H, Liu T, Chen X, Hao W, Fletcher PC (2011) Reduced dorsal prefrontal gray matter after chronic ketamine use. Biol Psychiatry 69:42–48

Lin PC, Lane HY, Lin CH (2016) Spontaneous remission of ketamine withdrawal-related depression. Clin Neuropharmacol 39:51–52

Liu Y, Lin D, Wu B, Zhou W (2016) Ketamine abuse potential and use disorder. Brain Res Bull 126:68–73

Livingston A, Waterman AE (1978) The development of tolerance to ketamine in rats and the significance of hepatic metabolism. Br J Pharmacol 64:63–69

Lua AC, Lin HR, Tseng YT, Hu AR, Yeh PC (2003) Profiles of urine samples from participants at rave party in Taiwan: prevalence of ketamine and MDMA abuse. Forensic Sci Int 136:47–51

Moore NN, Bostwick JM (1999) Ketamine dependence in anesthesia providers. Psychosomatics 40:356–359

Morgan CJ, Monaghan L, Curran HV (2004) Beyond the K-hole: a 3-year longitudinal investigation of the cognitive and subjective effects of ketamine in recreational users who have substantially reduced their use of the drug. Addiction 99:1450–1461

Morgan CJ, Rees H, Curran HV (2008) Attentional bias to incentive stimuli in frequent ketamine users. Psychol Med 38:1331–1340

Morgan CJ, Muetzelfeldt L, Curran HV (2009) Ketamine use, cognition and psychological wellbeing: a comparison of frequent, infrequent and ex-users with polydrug and non-using controls. Addiction 104:77–87

Morgan CJ, Muetzelfeldt L, Curran HV (2010) Consequences of chronic ketamine self-administration upon neurocognitive function and psychological wellbeing: a 1-year longitudinal study. Addiction 105:121–133

Morgan CJ, Curran HV, Independent Scientific Committee on Drugs (2012) Ketamine use: a review. Addiction 107:27–38

Muetzelfeldt L, Kamboj SK, Rees H, Taylor J, Morgan CJ, Curran HV (2008) Journey through the K-hole: phenomenological aspects of ketamine use. Drug Alcohol Depend 95:219–229

Murrough JW, Iosifescu DV, Chang LC, Al Jurdi RK, Green CE, Perez AM, Iqbal S, Pillemer S, Foulkes A, Shah A, Charney DS, Mathew SJ (2013) Antidepressant efficacy of ketamine in treatment-resistant major depression: a two-site randomized controlled trial. Am J Psychiatry 170:1134–1142

Ng SH, Tse ML, Ng HW, Lau FL (2010) Emergency department presentation of ketamine abusers in Hong Kong: a review of 233 cases. Hong Kong Med J 16:6–11

O'Brien CP (2005) Anticraving medications for relapse prevention: a possible new class of psychoactive medications. Am J Psychiatry 162:1423–1431

Olney JW, Wozniak DF, Jevtovic-Todorovic V, Farber NB, Bittigau P, Ikonomidou C (2002) Drug-induced apoptotic neurodegeneration in the developing brain. Brain Pathol 12:488–498

Pal HR, Berry N, Kumar R, Ray R (2002) Ketamine dependence. Anaesth Intensive Care 30:382–384

Petersen RC, Stillman RC (1978) Phencyclidine: an overview. NIDA Res Monogr:1–17

Rabiner EA (2007) Imaging of striatal dopamine release elicited with NMDA antagonists: is there anything there to be seen? J Psychopharmacol 21:253–258

Reier CE (1971) Ketamine—"dissociative agent" or hallucinogen? N Engl J Med 284:791–792

Rounsaville BJ (2004) Treatment of cocaine dependence and depression. Biol Psychiatry 56:803–809

Shahani R, Streutker C, Dickson B, Stewart RJ (2007) Ketamine-associated ulcerative cystitis: a new clinical entity. Urology 69:810–812

Smith GS, Schloesser R, Brodie JD, Dewey SL, Logan J, Vitkun SA, Simkowitz P, Hurley A, Cooper T, Volkow ND, Cancro R (1998) Glutamate modulation of dopamine measured in vivo

with positron emission tomography (PET) and 11C-raclopride in normal human subjects. Neuropsychopharmacology 18:18–25
Stone JM, Pepper F, Fam J, Furby H, Hughes E, Morgan C, Howes OD (2014) Glutamate, N-acetyl aspartate and psychotic symptoms in chronic ketamine users. Psychopharmacology (Berl) 231:2107–2116
Strong CE, Schoepfer KJ, Dossat AM, Saland SK, Wright KN, Kabbaj M (2017) Locomotor sensitization to intermittent ketamine administration is associated with nucleus accumbens plasticity in male and female rats. Neuropharmacology 121:195–203
Sun L, Li Q, Li Q, Zhang Y, Liu D, Jiang H, Pan F, Yew DT (2014) Chronic ketamine exposure induces permanent impairment of brain functions in adolescent cynomolgus monkeys. Addict Biol 19:185–194
Tang WK, Liang HJ, Lau CG, Tang A, Ungvari GS (2013) Relationship between cognitive impairment and depressive symptoms in current ketamine users. J Stud Alcohol Drugs 74:460–468
Tang J, Morgan HL, Liao Y, Corlett PR, Wang D, Li H, Tang Y, Chen J, Liu T, Hao W, Fletcher PC, Chen X (2015a) Chronic administration of ketamine mimics the perturbed sense of body ownership associated with schizophrenia. Psychopharmacology (Berl) 232:1515–1526
Tang WK, Morgan CJ, Lau GC, Liang HJ, Tang A, Ungvari GS (2015b) Psychiatric morbidity in ketamine users attending counselling and youth outreach services. Subst Abus 36:67–74
Trujillo KA, Smith ML, Sullivan B, Heller CY, Garcia C, Bates M (2011) The neurobehavioral pharmacology of ketamine: implications for drug abuse, addiction, and psychiatric disorders. ILAR J 52:366–378
UNODC (2019) World Drug Report United Nations publication
Venniro M, Mutti A, Chiamulera C (2015) Pharmacological and non-pharmacological factors that regulate the acquisition of ketamine self-administration in rats. Psychopharmacology (Berl) 232:4505–4514
Vollenweider FX, Leenders KL, Scharfetter C, Antonini A, Maguire P, Missimer J, Angst J (1997) Metabolic hyperfrontality and psychopathology in the ketamine model of psychosis using positron emission tomography (PET) and [18F]fluorodeoxyglucose (FDG). Eur Neuropsychopharmacol 7:9–24
Wang C, Zheng D, Xu J, Lam W, Yew DT (2013) Brain damages in ketamine addicts as revealed by magnetic resonance imaging. Front Neuroanat 7:23
Wang LJ, Chen CK, Lin SK, Chen YC, Xu K, Huang MC (2018) Cognitive profile of ketamine-dependent patients compared with methamphetamine-dependent patients and healthy controls. Psychopharmacology (Berl) 235:2113–2121
White PF, Way WL, Trevor AJ (1982) Ketamine—its pharmacology and therapeutic uses. Anesthesiology 56:119–136
Wong SW, Lee KF, Wong J, Ng WW, Cheung YS, Lai PB (2009) Dilated common bile ducts mimicking choledochal cysts in ketamine abusers. Hong Kong Med J 15:53–56
Xu K, Krystal JH, Ning Y, Chen Da C, He H, Wang D, Ke X, Zhang X, Ding Y, Liu Y, Gueorguieva R, Wang Z, Limoncelli D, Pietrzak RH, Petrakis IL, Zhang X, Fan N (2015) Preliminary analysis of positive and negative syndrome scale in ketamine-associated psychosis in comparison with schizophrenia. J Psychiatr Res 61:64–72
Yen CF, Hsu SY, Cheng CP (2007) Polysubstance use and its correlates in adolescent ecstasy users in Taiwan. Addict Behav 32:2286–2291
Yu WL, Cho CC, Lung PF, Hung EH, Hui JW, Chau HH, Chan AW, Ahuja AT (2014) Ketamine-related cholangiopathy: a retrospective study on clinical and imaging findings. Abdom Imaging 39:1241–1246
Zarate CA Jr, Singh JB, Carlson PJ, Brutsche NE, Ameli R, Luckenbaugh DA, Charney DS, Manji HK (2006) A randomized trial of an N-methyl-D-aspartate antagonist in treatment-resistant major depression. Arch Gen Psychiatry 63:856–864
Zhang Y, Xu Z, Zhang S, Desrosiers A, Schottenfeld RS, Chawarski MC (2014) Profiles of psychiatric symptoms among amphetamine type stimulant and ketamine using inpatients in Wuhan. China J Psychiatr Res 53:99–102

Brain Imaging of Ketamine Abusers

Yanhui Liao and Wei Hao

Abstract In this chapter, we highlight the role of brain imaging techniques (e.g., magnetic resonance imaging [MRI]) in studying ketamine abuse. In the past two to three decades, brain imaging studies demonstrated deficits in brain circuits related to drug addiction and drug abuse. This chapter begins with a brief introduction of structural and functional brain imaging techniques such as computed tomography (CT) and electroencephalogram (EEG). Then, we give a brief introduction of ketamine abuse in mainland China before introducing structural MRI and functional MRI and reviewing the application of structural MRI study for ketamine abusers (including reduction of gray matter volume and disruption of white matter integrity) and functional MRI study for ketamine abusers (including alternation of regional homogeneity (ReHo) of resting-state brain activity, functional connectivity by resting-state fMRI, task-based fMRI). Finally, we discuss the implication for medical use of ketamine by brain imaging study, especially its rapid-acting glutamatergic antidepressant effects and the "ketamine model" of psychosis.

Keywords Brain imaging · Magnetic resonance imaging · Structural MRI · Functional MRI · Chronic ketamine use · Ketamine abusers

Y. Liao (✉)
Department of Psychiatry, Sir Run Run Shaw Hospital, School of Medicine, Zhejiang University, Hangzhou, Zhejiang, China

Key Laboratory of Medical Neurobiology of Zhejiang Province, Hangzhou, Zhejiang, China

Department of Psychiatry, The Second Xiangya Hospital, Central South University, Changsha, Hunan, China

National Clinical Research Center on Mental Disorders, Changsha, Hunan, China

W. Hao
Department of Psychiatry, The Second Xiangya Hospital, Central South University, Changsha, Hunan, China

National Clinical Research Center on Mental Disorders, Changsha, Hunan, China

K. Hashimoto et al. (eds.), *Ketamine*,
https://doi.org/10.1007/978-981-15-2902-3_2

1 Brain Imaging Techniques

Brain imaging or neuroimaging techniques, by either directly or indirectly imaging the structure or function of the brain, have offered us a window to view noninvasively the human brain activity underlying complex affect, behavior, and cognition (Casey et al. 2005). Within the past two or three decades, by uncovering the inner workings of the human brain or revealing the neurological changes that result from brain diseases, these technological advances have accelerated the translation of basic neuroscience discoveries into precision medicine and targeted therapy in clinical management of brain diseases, which have made huge advances in cognitive and behavioral neuroscience.

Conceptually, brain imaging techniques can be classified into two broad categories: structural imaging and functional imaging (Hirsch et al. 2015). Structural imaging is used to reveal the anatomical properties of the brain, detect brain damage and abnormalities, and diagnose the gross intracranial disease and injury. Functional imaging is used to diagnose metabolic diseases and lesions on a finer scale or for neurological and cognitive psychology researches to identify brain areas and underlying brain processes that are linked to a particular cognitive or a behavioral task.

Here, we introduce some commonly used structural and functional brain imaging tools.

Structural brain imaging techniques:

1. Computed tomography (CT) or computed axial tomography (CAT) is a widely used approach that is typically used for viewing gross brain abnormalities, cerebral vascular accidents, and bone structure. However, it requires exposure to differential absorption of X-rays and has relatively low spatial resolution.
2. Magnetic resonance imaging (MRI) is a relatively accessible approach to visualize the soft tissues of brain such as gray matter and white matter with good spatial resolution. But it cannot be used in patients with metallic implants such as certain cardiac pacemakers or implantable cardioverter defibrillators.
3. Diffusion-based MRI (e.g., diffusion tensor imaging (DTI)) is an approach that is used for detecting detailed information of white matter integrity or microstructure and delineating white matter pathways that connect different regions of the brain. However, it requires complex image analyses and is sensitive to patient movement.

Functional brain imaging techniques:

1. Functional MRI (fMRI) and arterial spin labeling (ASL) can localize the brain activity associated with performing a cognitive task and/or behavior. It relies on the paramagnetic properties of oxygenated and deoxygenated hemoglobin to view images of blood flow changes associated with the regional activation of neurons during cognitive or behavioral tasks. The main drawback is its limited temporal resolution ability.

2. Positron emission tomography (PET) is an approach that is used to monitor the brain activity as well as the metabolism associated with performing a cognitive task and/or behavior by measuring emissions from radioactively labeled, metabolically active chemicals (e.g., fludeoxyglucose F 18) that have been injected into the bloodstream. However, it requires the injection of radioactive tracers, which is a main drawback.
3. Electroencephalogram (EEG) and evoked related potentials (ERP) can directly record the underlying electrical brain activity that is associated with a cognitive task and/or behavior with good temporal resolution. However, these techniques have poor spatial resolution compared with fMRI and require a complicated analysis of the acquired data.

Many other brain imaging techniques are also available, such as the functional near-infrared spectroscopy (fNIRS) and magnetoencephalogram (MEG). In this chapter, we focus on MRI (structural and functional) techniques because these are most widely used techniques in drug abuse and addiction (Parvaz et al. 2011). This chapter mainly focuses on structural and functional brain imaging of ketamine abusers. We briefly introduce ketamine abuse in mainland China before reviewing neuroimaging studies of ketamine addiction and abuse.

2 Nonmedical Use of Ketamine in Mainland China

During the last three or four decades, drug abuse spread quickly following its reemergence as a national problem in China (Liu et al. 2016). The number of registered drug abusers increased from 70,000 in 1990 to more than one million by the end of 2005 (Michels et al. 2007). According to China drug situation report 2018, the number of registered drug abusers doubled in the next decade to more than two million by the end of 2018, and the number of abusers of "new" types of drugs including methamphetamine (accounting for 56.1%) and ketamine (accounting for 2.6%) exceeded the number of abusers of opioids (www.nncc626.com). Illicit drug trafficking and production have swept in mainland China, particularly in southern China, which have caused many problems (such as the spread of HIV) for both abusers and the community (Fang et al. 2006).

In mainland China, ketamine was rescheduled from Schedule II in 2001 to Schedule I in 2007 (Liao et al. 2017). It is commonly abused through snorting. Figure 1 shows a typical way of making and snorting ketamine powder. Long-term ketamine abuse often results in bladder dysfunction (Tsai et al. 2009), sleeping disturbances (Tang et al. 2015a), cognitive impairment (Chan et al. 2013; Morgan et al. 2004a, b; Tang et al. 2013), depression (Li et al. 2017a; Tang et al. 2013), and even schizophrenia-like symptoms (Liao et al. 2016a; Tang et al. 2015b). Furthermore, a line of study reported brain structural and functional alterations in ketamine abusers (Liao et al. 2010, 2011, 2012, 2016b, 2018; Wang et al. 2013). This chapter reviews

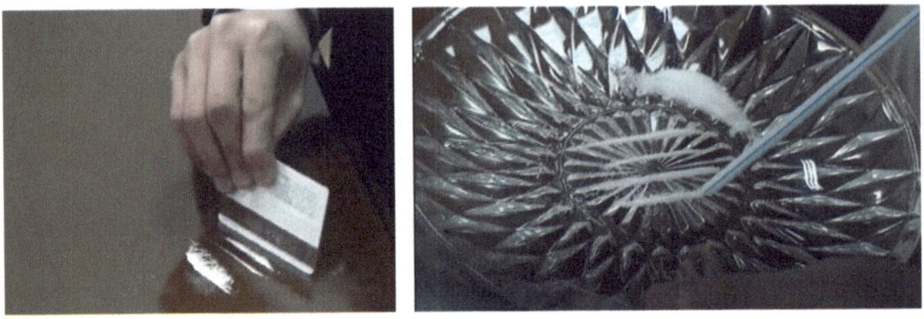

Fig. 1 Typical way of preparing (left) and snorting (right) ketamine powder

the available structural and functional neuroimaging (mainly MRI) studies of ketamine abusers in detail.

3 Brain Imaging of Ketamine Abusers

Magnetic resonance imaging (MRI) is one of the most commonly used techniques to examine the brain activity in healthy brain and brain with disease, such as addiction or other mental illness. It is often divided into structural MRI and functional MRI (fMRI) (Symms et al. 2004). The field strength of the magnet is measured in tesla (T), and the magnetic strength for most clinical or research MRI scanners is 1.5 or 3 T. Most studies indicate that 3.0-T MRI scanner performs as good as or better than 1.5-T MRI scanner (Wood et al. 2012).

3.1 Magnetic Resonance Imaging

In the magnetic field, the nuclear spins of certain atoms within an object are oriented either parallel or antiparallel to the main magnetic field and produce a secondary spin or wobble (precession) of nuclei around the main or static magnetic field with a certain frequency called the Larmor frequency (named after the Irish physicist and mathematician Joseph Larmor, 1857–1942). The Larmor or precessional frequency in MRI is related to the strength of the magnetic field (the B0 field). Magnetic resonance occurs when a radio frequency (RF) pulse produced at the (tissue specific) Larmor frequency, exciting and then raising the nuclear spin orientation of hydrogen atoms from lower to higher energy states. By rotating the magnetization, the RF field is switched off and the magnetization once again freely precesses around the direction of the original magnetization. This exchange of energy between spin states is called resonance, and a computer displays the different resonance characteristics of various tissue types as an image. This time-dependent precession produces a current in the receiver RF coil, and the exponentially decaying resultant current (the free induction decay) constitutes the MR signal. The amount of the signal that is used to compose an image is proportional to the magnetic field strength of the scanner. (Signal-to-noise ratio (SNR), contrast-to-noise ratio (CNR), spatial resolution, and temporal resolution are commonly used technical terms to describe MRI image quality.) During the energy transition period, magnetization returns to its original equilibrium state (relaxation), characterized by T1 and T2 time constants, which depend on physical and chemical characteristics unique to the tissue type. The T1 and T2 differences between different tissue types (e.g., gray matter and white matter) produce a high-contrast MR image (Lauterbur 1973; Mansfield and Maudsley 1977).

3.1.1 Structural MRI

Structural MRI is a noninvasive technique that is used for examining the anatomy and pathology of the brain (as opposed to fMRI). It produces images that can be used for clinical radiological reporting as well as for diagnosis or research with detailed analysis. Structural MRI includes high-resolution imaging, T2 relaxation measurement, T2-weighted imaging, T1 relaxation measurement, magnetization transfer imaging, and diffusion imaging (Symms et al. 2004). Some MRI scan sequences are volumetric, and the volumes of regional gray matter and white matter often change considerably in individuals with brain diseases.

3.1.2 Functional MRI

Since its inception in 1990, functional MRI (fMRI) has been used to map human brain function noninvasively and rapidly with full brain coverage, and with relatively high spatial and temporal resolution. John Belliveau, Bruce Rosen et al. introduced it by using gadolinium as a contrast agent (Belliveau et al. 1990; Rosen et al. 1990). This was then immediately followed by a series of fMRI studies using the "blood oxygen level dependent" (BOLD) signal (Hoge et al. 1999; Kwong et al. 1992; Ogawa et al. 1990), which is the best known oxygen sensitive contrast agent. BOLD contrast results from a change in the magnetic field surrounding the red blood cells depending on the oxygen state of the protein hemoglobin, which is used for the indirect measurement of the brain activity.

In the past three decades, with extraordinary features (e.g., widespread availability, noninvasive nature, relatively low cost, and good spatial resolution), fMRI was applied in an exceptionally large number of studies in the cognitive neurosciences, clinical psychiatry/psychology, and presurgical planning to find biomarkers for disease (Bush et al. 1999; Fleisher et al. 2009), to monitor therapy (Stoeckel et al. 2014), or to study pharmacological efficacy (Honey and Bullmore 2004). The use of fMRI at rest has also enabled researchers to investigate resting functional connectivity of the human brain (Biswal et al. 1995). Measures of resting functional connectivity have been shown to be sensitive to brain diseases including drug and nondrug addiction (Fedota and Stein 2015; Hong et al. 2013; Sutherland et al. 2012). The limitations of fMRI include high susceptibility of the BOLD response to several nonneural and imaging artifacts, especially due to its low SNR and low temporal resolution.

By comparing brain structure, function, and metabolism between drug-abusing and non-abusing individuals, brain imaging techniques enable researchers to observe the disruption of ketamine or other drug abusers' brain activity and allows us to better understand the mechanisms of addiction (Volkow et al. 2014).

3.2 Structural MRI Study of Ketamine Abusers

In the past two decades, there was considerable evidence that many psychiatric disorders (e.g., schizophrenia (Canu et al. 2015; Van Erp et al. 2018), bipolar disorder (Hibar et al. 2016; Strakowski et al. 1999), major depressive disorder (Schmaal et al. 2017), autism spectrum disorder (Sparks et al. 2002), obsessive–compulsive disorder (Fouche et al. 2017; Jenike et al. 1996), and attention-deficit/hyperactivity disorder (Seidman et al. 2005)) are associated with either increased or decreased regional brain volumes compared with gender- and age-matched healthy subjects. These brain structural abnormalities were also reported among ketamine abusers. Until now, only a few clinical structural MRI findings, however, have reported that the long-term effects of ketamine abuse is associated with the reduction of gray matter volume and the disruption of white mater integrity.

to assess regional gray matter volume reduction in ketamine abusers, voxel-based morphometry in conjunction with statistical parametric mapping on the structural magnetic resonance images of 41 ketamine abusers and 44 drug-naive control individuals in mainland China was used. This study found reduction of dorsal prefrontal gray matter after long-term repeated ketamine abuse. In particular, this study revealed significant reduction in gray matter volume in left superior frontal gyrus and right middle frontal gyrus of ketamine abusers in comparison with age-matched healthy volunteers (see Fig. 2). Furthermore, the duration of ketamine abuse was negatively correlated with the reduction of gray matter volume in bilateral frontal cortex, whereas the estimated total lifetime ketamine consumption was negatively correlated with the reduction of gray matter volume in left superior frontal gyrus. The association between frontal gray matter loss and the duration of ketamine use or cumulative doses of ketamine may suggest a dose-dependent effect of long-term abuse of the drug (Liao et al. 2011). Neuroimaging studies in other drug addictive users have also identified dysfunction of the prefrontal cortex, which suggests the involvement of this brain region in drug addiction (Goldstein and Volkow 2011).

Several studies, mainly in China and the UK, uncovered the disruption of white matter in chronic ketamine abusers. A study with the same Chinese sample (41 ketamine-dependent subjects and 44 drug-free healthy volunteers), using in vivo diffusion tensor magnetic resonance imaging data, measured white matter volumes. This study also found bilateral frontal and left temporoparietal white matter

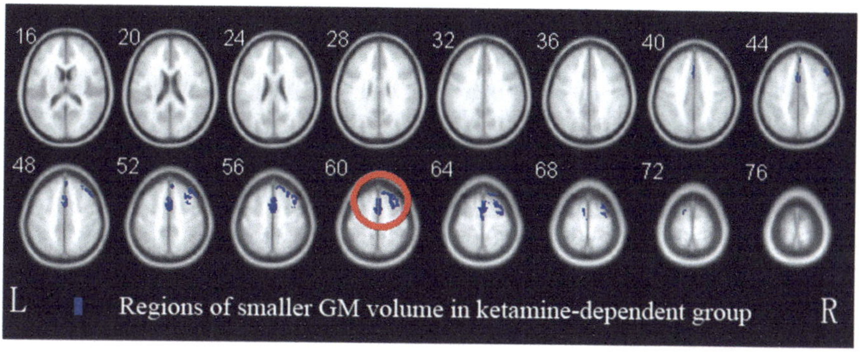

Fig. 2 Dorsal prefrontal gray matter reduction in ketamine abuses compared with healthy controls

alterations associated with long-term ketamine abuse (see Fig. 3). Furthermore, frontal white matter fractional anisotropy (FA is a scalar value between 0 and 1 that describes the degree of anisotropy of a diffusion process.) correlated with the severity of drug use (as measured by estimating totally abused ketamine). This study provided first evidence for dose-dependent abnormalities of white matter in bilateral frontal and left temporoparietal regions following long-term ketamine abuse. These findings further suggest a microstructural basis for the changes in cognition and experience observed with chronic ketamine abuse (Liao et al. 2010).

Another study in the UK measured indices of white matter microstructural integrity and connectivity in 16 ketamine abusers and 16 poly-drug-using controls. The study used probabilistic tractography to quantify alterations in corticosubcortical connectivity associated with repeated ketamine abuse. This study found a decrease in the axial diffusivity profile of white matter in the right hemisphere network of white matter regions in ketamine abusers compared with poly-drug abusers. For ketamine abusers, this study found that the frequency of dissociative experiences reported by chronic ketamine users associated positively with the changes of connectivity in the caudate nucleus and the lateral prefrontal cortex. These findings

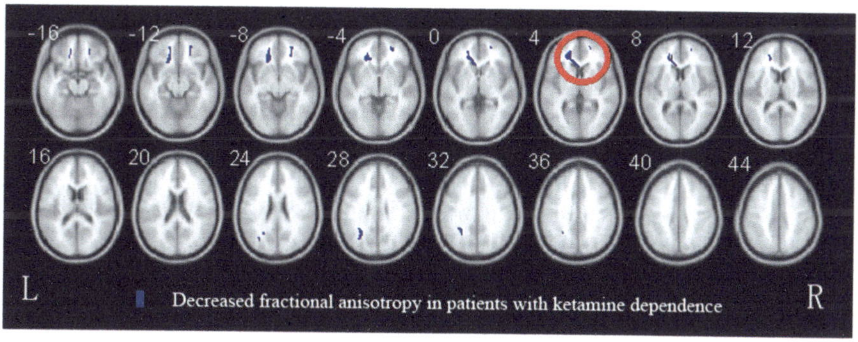

Fig. 3 Frontal white matter alterations in ketamine abuses compared with healthy controls

further confirmed that ketamine abuse may be associated with widespread alternation of white matter integrity. It also suggests that white matter pathways between subcortical and prefrontal cortical regions may in part predict individual differences in the frequency of dissociative experiences due to ketamine abuse (Roberts et al. 2014).

By employing MRI, a study from the south of China illustrated the possible disrupted brain regions due to ketamine abuse by comparing 21 ketamine addicts with three age-matched healthy controls. This study revealed the lesions in many regions (e.g., prefrontal, parietal, occipital, limbic, brainstem, and corpus striatum) of the brain of ketamine abusers (Wang et al. 2013).

Several animal studies using monkeys or rats (Li et al. 2017b; Sun et al. 2014; Yeung et al. 2010; Zou et al. 2009) have also shown that brain structural, especially frontal cortex, disruption is associated with long-term exposure to ketamine.

3.3 *Functional MRI Study of Ketamine Abusers*

Only a few studies measured the long-term effects of ketamine abuse on brain functional alternations. One resting-state fMRI demonstrated the alterations in regional homogeneity (ReHo) of resting-state brain activity in long-term ketamine abusers. This study examined such effects on spontaneous brain dynamics between 41 patients with ketamine dependence and 44 healthy control subjects by resting-state fMRI. The results showed that compared with healthy controls, ketamine abusers displayed decreased ReHo in the right anterior cingulate cortex and increased ReHo in left precentral frontal gyrus. This study also showed negative correlations between increased ReHo in precentral frontal gyrus and estimated totally abused ketamine in their lifetime and levels of ketamine craving. The findings from this study indicate that ketamine abuse might be associated with alterations in the functional connectivity of medial and lateral prefrontal cortices (Liao et al. 2012).

Decreased thalamocortical connectivity also has been observed in chronic ketamine abusers, which may suggest its role in the mechanistic "switch" from recreational abuse to dysregulated ketamine addiction. In this study, 41 ketamine abusers and 89 healthy control subjects were enrolled to measure the functional connectivity within the cerebral cortex by resting-state fMRI. This study found that ketamine abusers showed significantly decreased (but not increased) connectivity between the thalamic nuclear groups and the cortical regions of interest (ROI), including the prefrontal cortex, the motor cortex/supplementary motor area, and the posterior parietal cortex (see Fig. 4). This finding provided the first evidence of abnormal thalamocortical connectivity of resting state brain activity in long-term ketamine abusers (Liao et al. 2016b).

Ketamine abuse-related cues may induce craving in ketamine-dependent patients. A study measured regional brain activation to ketamine, cigarette smoking, and sexual activity. This functional MRI study recruited 40 smokers with ketamine abuse, 45 smokers without ketamine abuse, and 44 non-ketamine abuse noncigarette

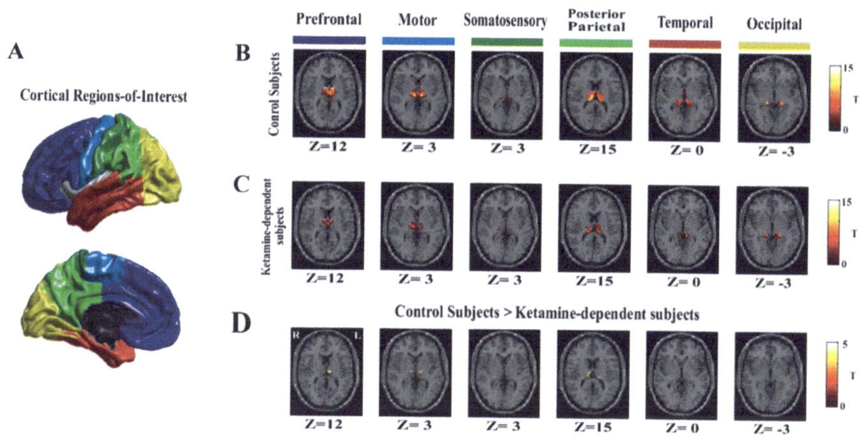

Fig. 4 The diffusively decreased thalamocortical connectivity in ketamine-dependent subjects. (**a**) The six cortical regions of interest. (**b**) Thalamocortical connectivity of the drug-free control subjects. (**c**) Thalamocortical connectivity of the ketamine-dependent subjects. (**d**) Group differences in thalamocortical connectivity for the drug-free control subjects and the ketamine-dependent subjects

smoking healthy controls to view ketamine use-related cigarette smoking and sexual films. This task-based fMRI found that smokers with ketamine abuse showed significant increased activation in regions of anterior cingulate cortex and precuneus in response to ketamine cues. They also showed reduced activation in cerebellum and middle temporal cortex in response to sexual cues. Smokers (both with and without ketamine abuse) showed increased activation in the right precentral frontal cortex in response to smoking cues. Non-ketamine users (both smokers without ketamine abuse and controls) showed increased activation of cerebellum and middle temporal cortex when they were viewing sexual cues (see Fig. 4). These findings indicated the engagement of distinct neural circuitry for ketamine-related stimuli (compared with cigarette smoking stimuli or sexual stimuli) in ketamine abusers. While smokers (for both with and without ketamine abuse) showed overlapping differences in activation for smoking cues (Liao et al. 2018), which may be partly due to the interaction effects of ketamine and nicotine in multiple brain regions. Nicotine substantially ameliorated the effects of ketamine on anterior cingulate regional cerebral blood flow (rCBF), and those effects may link to psychosis, reward, and addictive behaviors (Rowland et al. 2010).

Besides measuring functional MRI, the consequences of ketamine abuse in the human brain function have also been studied using positron emission tomography (PET) in a group of 14 recreational chronic ketamine users and matched healthy subjects. The study found that chronic ketamine abuse exhibited a regionally selective upregulation of D1 receptor availability in the dorsolateral prefrontal cortex (Narendran et al. 2005). Moreover, functional MRI of the brains in adolescent monkeys with chronic exposure to ketamine showed regional brain functional changes, particularly in the prefrontal dopaminergic system – a system critically involved in working memory and executive function (Yu et al. 2012). Thus, long-term ketamine administration may involve neurodegenerative process similar to that of aging and/or Alzheimer's disease, which was demonstrated in both animal (Yeung et al. 2010) and human (Chan et al. 2013) studies.

Table 1 presents a summary of regional brain structural and functional disruption in ketamine abusers.

4 Implication for Medical Use of Ketamine by Brain Imaging Study

In addition to its abuse potential, ketamine is commonly used for medical purposes (Morgan et al. 2012). Ketamine is a noncompetitive antagonist at the glutamatergic *N*-methyl-D-aspartate (NMDA) receptor that is most widely used injectable anesthetic agent in veterinary medicine (Clarke and Hall 1990; Young et al. 1993). It is also used in specialist anesthesia, particularly pediatrics and field medicine for human (Cartwright and Pingel 1984; Green and Johnson 1990). Furthermore, ketamine has a role in pain management in both human and veterinary medicine (Elia and Tramèr 2005). Ketamine, as an *N*-methyl-D-aspartate (NMDA) receptor antagonist, has been found to induce schizophrenia-type symptoms in humans. Thus, there

Table 1 Regional brain structural and functional disruption in ketamine abusers

	Sample	Disrupted brain regions
Structural MRI	41 ketamine abusers 44 age-matched healthy controls	Dorsal prefrontal gray matter reduction (Liao et al. 2011)
	41 ketamine abusers 44 age-matched healthy controls	Bilateral frontal and left temporoparietal white matter alterations (Liao et al. 2010)
	16 ketamine abusers 16 poly-drug abusers	Reduction in the axial diffusivity profile of white matter in a right hemisphere network of white matter regions (Roberts et al. 2014)
	21 ketamine addicts 3 age-matched normal subjects	Brain damages in prefrontal, parietal, occipital, limbic, brainstem, and corpus striatum (Wang et al. 2013)
Functional MRI	41 ketamine abusers 44 age-matched healthy controls	Alterations in regional homogeneity of resting-state brain activity in ketamine addicts (Liao et al. 2012)
	41 ketamine abusers 89 healthy control subjects	Decreased thalamocortical connectivity in chronic ketamine users (Liao et al. 2016b)
	40 smokers with ketamine abuse 45 smokers without ketamine abuse 44 healthy controls	Cue-induced brain activation in chronic ketamine-dependent subjects, cigarette smokers, and healthy controls: a task functional magnetic resonance imaging (Liao et al. 2018)

are also a line of experimental studies using brain imaging to explore the "ketamine model" of psychosis (Deakin et al. 2008; Honey et al. 2008). As ketamine can block bursting in the lateral habenula to rapidly relieve depressive symptoms (Yang et al. 2018), it is used today for the treatment of depression, particularly for treatment-resistant depression (Aan Het Rot et al. 2012; Katalinic et al. 2013; Rosenblat et al. 2019). Recently, it was also found to be promising for addiction relapse prevention (e.g., heroin and alcohol addiction) as it reduced symptoms of depression (Ezquerra-Romano et al. 2018).

A series of pharmacological MRI study examined ketamine's pharmacological effects on brain function, mainly to explore its potent and fast-acting antidepressant effects and uncover the neural mechanisms of psychosis and schizophrenia (as it produces similar symptoms).

4.1 For Depression

The abuse potential and psychotomimetic effects of ketamine may be linked with the dopaminergic system that also associated with antidepressant effects. Traditional antidepressants require several weeks to show their therapeutic effects. A single subanesthetic dose of the *N*-methyl-D-aspartate glutamate receptor antagonist ketamine, however, can produce significant clinical improvement within hours, getting the approval of Food and Drug Administration (FDA) of esketamine nasal spray for patients with treatment-resistant depression. Ketamine as a rapid-acting glutamatergic antidepressant, its novel treatment mechanisms are emerging a line of research (Krystal et al. 2013). The antidepressant properties of (*R*,*S*)-ketamine have also relevance for the development of next-generation, rapid-acting antidepressants (Zanos et al. 2016).

Subanesthetic doses of ketamine increase dopamine levels in the frontal cortex, in the striatum, and in the nucleus accumbens in rodents (Kokkinou et al. 2017). Applying 1H-MRS to 10 major depressive disorder (MDD) patients, however, did not observe the association between ketamine infusion with changes in occipital amino acid neurotransmitter (AANt) content, which indicates that these changes are not a correlate of ketamine's antidepressant action (Valentine et al. 2011). A study using resting-state fMRI to investigate the intrinsic brain networks in MDD patients with ketamine treatment showed that ketamine significantly increased global signal regression (GBCr) in the PFC and reduced GBCr in the cerebellum, which may serve as a putative marker for successful treatment and a target for antidepressants' development (Abdallah et al. 2017). However, it is important to be aware of ketamine abuse and addiction potential, even in MDD patients who received ketamine for antidepressant purposes (Bonnet 2015). In the future, research should explore more details about ketamine's acute and chronic antidepressant effects on neuropharmacological and cognitive levels.

4.2 For Psychosis and Schizophrenia

On the contrary, several studies indicate that subanesthetic doses of ketamine impair prefrontal cortex (PFC) function in the rat and produce schizophrenia-like symptoms and dissociative states, including impaired performance of frontal lobe-sensitive tests, which is similar to cognitive impairment (Morgan and Curran 2006), persistent dissociative, depressive, and delusional thinking (Sassano-Higgins et al. 2016) by long-term ketamine use. It is suggested that ketamine may disrupt PFC function partly by interacting with dopaminergic neurotransmission in this region. By increasing the release of glutamate, thereby stimulating postsynaptic non-NMDA glutamate receptors, it probably impairs cognitive functions associated with PFC as a result (Moghaddam et al. 1997). Intravenous ketamine (1-min bolus of 0.26 mg/kg) to healthy individuals induced a rapid, focal, and unexpected decrease of the brain activity in ventromedial frontal cortex, including orbitofrontal cortex and subgenual cingulate, which strongly predicted its dissociative effects and increased activity in regions of midposterior cingulate, thalamus, and temporal cortex (Deakin et al. 2008).

Human and animal studies indicate that NMDA receptor hypofunction has been implicated in the pathophysiology of psychosis and schizophrenia and diminishes the inhibitory control of PFC output neurons. The findings of the present study suggest that both chronic ketamine users (Narendran et al. 2005) and schizophrenia patients (Laruelle et al. 2003) display the same endophenotypic trait – upregulated D1 receptor expression in the dorsolateral PFC, which supports the hypothesis that this alteration might be secondary to NMDA dysfunction in schizophrenia. For example, a 4-T 1H proton magnetic resonance spectroscopy (1H-MRS) study with ketamine administration in 10 healthy subjects found increased glutamatergic activity in the anterior cingulate, which may associate with acute hypofunctional NMDA receptor state (Rowland et al. 2005). Animal study also found its NMDA receptor hypofunction. Thus, reducing this effect may be critical for the treatment of psychosis and schizophrenia (Homayoun and Moghaddam 2007).

5 Conclusion

Brain imaging technology has had a tremendous impact on evaluating the neural mechanism associated with acute or chronic administration of ketamine. It provided a basic knowledge of ketamine's medical use- or abuse-related brain circuits and the related cognitive and behavior outcomes. However, a major caveat remains in the uncertainty is its long-term safety, such as how to use ketamine for MDD treatment while limiting its psychotomimetic and dissociative side effects as well as addiction and abuse potential. Novel analytical approaches for brain imaging data should also be developed to facilitate the translation of findings from the research to the clinical setting, promoting the medical benefits of ketamine's medical use while preventing from its addiction and abuse potential. In the future, the insights provided by neuroimaging studies of ketamine could contribute to biomarker development for its effective treatment of MDD and even other disorders, and to new approaches to discover better treatment strategies.

References

Aan Het Rot M, Zarate CA Jr, Charney DS, Mathew SJ (2012) Ketamine for depression: where do we go from here? Biol Psychiatry 72:537–547

Abdallah CG, Averill LA, Collins KA, Geha P, Schwartz J, Averill C, DeWilde KE, Wong E, Anticevic A, Tang CY (2017) Ketamine treatment and global brain connectivity in major depression. Neuropsychopharmacology 42:1210

Belliveau JW, Rosen BR, Kantor HL, Rzedzian RR, Kennedy DN, McKinstry RC, Vevea JM, Cohen MS, Pykett IL, Brady TJ (1990) Functional cerebral imaging by susceptibility-contrast NMR. Magn Reson Med 14:538–546

Biswal B, Zerrin Yetkin F, Haughton VM, Hyde JS (1995) Functional connectivity in the motor cortex of resting human brain using echo-planar MRI. Magn Reson Med 34:537–541

Bonnet U (2015) Long-term ketamine self-injections in major depressive disorder: focus on tolerance in ketamine's antidepressant response and the development of ketamine addiction. J Psychoactive Drugs 47:276–285

Bush G, Frazier JA, Rauch SL, Seidman LJ, Whalen PJ, Jenike MA, Rosen BR, Biederman J (1999) Anterior cingulate cortex dysfunction in attention-deficit/hyperactivity disorder revealed by fMRI and the Counting Stroop. Biol Psychiatry 45:1542–1552

Canu E, Agosta F, Filippi M (2015) A selective review of structural connectivity abnormalities of schizophrenic patients at different stages of the disease. Schizophr Res 161:19–28

Cartwright P, Pingel S (1984) Midazolam and diazepam in ketamine anaesthesia. Anaesthesia 39:439–442

Casey BJ, Tottenham N, Liston C, Durston S (2005) Imaging the developing brain: what have we learned about cognitive development? Trends Cogn Sci 9:104–110

Chan KW, Lee TM, Siu AM, Wong DP, Kam C-M, Tsang SK, Chan CC (2013) Effects of chronic ketamine use on frontal and medial temporal cognition. Addict Behav 38:2128–2132

Clarke K, Hall L (1990) A survey of anaesthesia in small animal practice: AVA/BSAVA report. J Assoc Vet Anaesth Great Britain Ireland 17:4–10

Deakin JW, Lees J, McKie S, Hallak JE, Williams SR, Dursun SM (2008) Glutamate and the neural basis of the subjective effects of ketamine: a pharmaco–magnetic resonance imaging study. Arch Gen Psychiatry 65:154–164

Elia N, Tramèr MR (2005) Ketamine and postoperative pain—a quantitative systematic review of randomised trials. Pain 113:61–70

Ezquerra-Romano II, Lawn W, Krupitsky E, Morgan C (2018) Ketamine for the treatment of addiction: evidence and potential mechanisms. Neuropharmacology 142:72–82

Fang YX, Wang YB, Shi J, Liu ZM, Lu L (2006) Recent trends in drug abuse in China. Acta Pharmacol Sin 27:140–144

Fedota JR, Stein EA (2015) Resting-state functional connectivity and nicotine addiction: prospects for biomarker development. Ann N Y Acad Sci 1349:64

Fleisher AS, Sherzai A, Taylor C, Langbaum JB, Chen K, Buxton RB (2009) Resting-state BOLD networks versus task-associated functional MRI for distinguishing Alzheimer's disease risk groups. Neuroimage 47:1678–1690

Fouche J-P, Du Plessis S, Hattingh C, Roos A, Lochner C, Soriano-Mas C, Sato JR, Nakamae T, Nishida S, Kwon JS (2017) Cortical thickness in obsessive–compulsive disorder: multisite mega-analysis of 780 brain scans from six centres. Br J Psychiatry 210:67–74

Goldstein RZ, Volkow ND (2011) Dysfunction of the prefrontal cortex in addiction: neuroimaging findings and clinical implications. Nat Rev Neurosci 12:652

Green SM, Johnson NE (1990) Ketamine sedation for pediatric procedures: part 2, review and implications. Ann Emerg Med 19:1033–1046

Hibar D, Westlye LT, van Erp TG, Rasmussen J, Leonardo CD, Faskowitz J, Haukvik UK, Hartberg CB, Doan NT, Agartz I (2016) Subcortical volumetric abnormalities in bipolar disorder. Mol Psychiatry 21:1710

Hirsch GV, Bauer CM, Merabet LB (2015) Using structural and functional brain imaging to uncover how the brain adapts to blindness. Ann Neurosci Psychol 2:5

Hoge RD, Atkinson J, Gill B, Crelier GR, Marrett S, Pike GB (1999) Investigation of BOLD signal dependence on cerebral blood flow and oxygen consumption: the deoxyhemoglobin dilution model. Magn Reson Med 42:849–863

Homayoun H, Moghaddam B (2007) NMDA receptor hypofunction produces opposite effects on prefrontal cortex interneurons and pyramidal neurons. J Neurosci 27:11496–11500

Honey G, Bullmore E (2004) Human pharmacological MRI. Trends Pharmacol Sci 25:366–374

Honey GD, Corlett PR, Absalom AR, Lee M, Pomarol-Clotet E, Murray GK, McKenna PJ, Bullmore ET, Menon DK, Fletcher PC (2008) Individual differences in psychotic effects of ketamine are predicted by brain function measured under placebo. J Neurosci 28:6295–6303

Hong S-B, Zalesky A, Cocchi L, Fornito A, Choi E-J, Kim H-H, Suh J-E, Kim C-D, Kim J-W, Yi S-H (2013) Decreased functional brain connectivity in adolescents with internet addiction. PLoS One 8:e57831

Jenike MA, Breiter HC, Baer L, Kennedy DN, Savage CR, Olivares MJ, O'Sullivan RL, Shera DM, Rauch SL, Keuthen N (1996) Cerebral structural abnormalities in obsessive-compulsive disorder: a quantitative morphometric magnetic resonance imaging study. Arch Gen Psychiatry 53:625–632

Katalinic N, Lai R, Somogyi A, Mitchell PB, Glue P, Loo CK (2013) Ketamine as a new treatment for depression: a review of its efficacy and adverse effects. Aust N Z J Psychiatry 47:710–727

Kokkinou M, Ashok AH, Howes OD (2017) The effects of ketamine on dopaminergic function: meta-analysis and review of the implications for neuropsychiatric disorders. Mol Psychiatry 23:59

Krystal JH, Sanacora G, Duman RS (2013) Rapid-acting glutamatergic antidepressants: the path to ketamine and beyond. Biol Psychiatry 73:1133–1141

Kwong KK, Belliveau JW, Chesler DA, Goldberg IE, Weisskoff RM, Poncelet BP, Kennedy DN, Hoppel BE, Cohen MS, Turner R (1992) Dynamic magnetic resonance imaging of human brain activity during primary sensory stimulation. Proc Natl Acad Sci 89:5675–5679

Laruelle M, Kegeles LS, Abi-Dargham A (2003) Glutamate, dopamine, and schizophrenia. Ann N Y Acad Sci 1003:138–158

Lauterbur PC (1973) Image formation by induced local interactions: examples employing nuclear magnetic resonance. Nature 242:190–191

Li C-SR, Zhang S, Hung C-C, Chen C-M, Duann J-R, Lin C-P, Lee TS-H (2017a) Depression in chronic ketamine users: sex differences and neural bases. Psychiatry Res Neuroimaging 269:1–8

Li Q, Shi L, Lu G, Yu H-L, Yeung F-K, Wong N-K, Sun L, Liu K, Yew D, Pan F (2017b) Chronic ketamine exposure causes white matter microstructural abnormalities in adolescent cynomolgus monkeys. Front Neurosci 11:285

Liao Y, Tang J, Ma M, Wu Z, Yang M, Wang X, Liu T, Chen X, Fletcher PC, Hao W (2010) Frontal white matter abnormalities following chronic ketamine use: a diffusion tensor imaging study. Brain 133:2115–2122

Liao Y, Tang J, Corlett PR, Wang X, Yang M, Chen H, Liu T, Chen X, Hao W, Fletcher PC (2011) Reduced dorsal prefrontal gray matter after chronic ketamine use. Biol Psychiatry 69:42–48

Liao Y, Tang J, Fornito A, Liu T, Chen X, Chen H, Xiang X, Wang X, Hao W (2012) Alterations in regional homogeneity of resting-state brain activity in ketamine addicts. Neurosci Lett 522:36–40

Liao Y, Qi C, Wu Q, Tang J (2016a) Psychiatric symptoms in individuals who use ketamine versus methamphetamine—implications for glutamatergic and dopaminergic model for schizophrenia: a cohort study. Lancet 388:S67
Liao Y, Tang J, Liu J, Xie A, Yang M, Johnson M, Wang X, Deng Q, Chen H, Xiang X (2016b) Decreased thalamocortical connectivity in chronic ketamine users. PLoS One 11:e0167381
Liao Y, Tang Y-l, Hao W (2017) Ketamine and international regulations. Am J Drug Alcohol Abuse 43:495–504
Liao Y, Johnson M, Qi C, Wu Q, Xie A, Liu J, Yang M, Huang M, Zhang Y, Liu T (2018) Cue-induced brain activation in chronic ketamine-dependent subjects, cigarette smokers, and healthy controls: a task functional magnetic resonance imaging study. Front Psych 9:88
Liu Y, Lin D, Wu B, Zhou W (2016) Ketamine abuse potential and use disorder. Brain Res Bull 126:68–73
Mansfield P, Maudsley AA (1977) Medical imaging by NMR. Br J Radiol 50:188–194
Michels II, Fang Y-x, Zhao D, Zhao L-y, Lu L (2007) Comparison of drug abuse in Germany and China. Acta Pharmacol Sin 28:1505
Moghaddam B, Adams B, Verma A, Daly D (1997) Activation of glutamatergic neurotransmission by ketamine: a novel step in the pathway from NMDA receptor blockade to dopaminergic and cognitive disruptions associated with the prefrontal cortex. J Neurosci 17:2921–2927
Morgan CJ, Curran HV (2006) Acute and chronic effects of ketamine upon human memory: a review. Psychopharmacology (Berl) 188:408–424
Morgan CJ, Monaghan L, Curran HV (2004a) Beyond the K-hole: a 3-year longitudinal investigation of the cognitive and subjective effects of ketamine in recreational users who have substantially reduced their use of the drug. Addiction 99:1450–1461
Morgan CJ, Riccelli M, Maitland CH, Curran HV (2004b) Long-term effects of ketamine: evidence for a persisting impairment of source memory in recreational users. Drug Alcohol Depend 75:301–308
Morgan CJ, Curran HV, Drugs ISC o (2012) Ketamine use: a review. Addiction 107:27–38
Narendran R, Frankle WG, Keefe R, Gil R, Martinez D, Slifstein M, Kegeles LS, Talbot PS, Huang Y, Hwang D-R (2005) Altered prefrontal dopaminergic function in chronic recreational ketamine users. Am J Psychiatry 162:2352–2359
Ogawa S, Lee TM, Kay AR, Tank DW (1990) Brain magnetic resonance imaging with contrast dependent on blood oxygenation. Proc Natl Acad Sci U S A 87:9868–9872
Parvaz MA, Alia-Klein N, Woicik PA, Volkow ND, Goldstein RZ (2011) Neuroimaging for drug addiction and related behaviors. Rev Neurosci 22:609–624
Roberts RE, Curran HV, Friston KJ, Morgan CJ (2014) Abnormalities in white matter microstructure associated with chronic ketamine use. Neuropsychopharmacology 39:329
Rosen BR, Belliveau JW, Vevea JM, Brady TJ (1990) Perfusion imaging with NMR contrast agents. Magn Reson Med 14:249–265
Rosenblat JD, Carvalho AF, Li M, Lee Y, Subramanieapillai M, McIntyre RS (2019) Oral ketamine for depression: a systematic review. J Clin Psychiatry 80:18r12475
Rowland LM, Bustillo JR, Mullins PG, Jung RE, Lenroot R, Landgraf E, Barrow R, Yeo R, Lauriello J, Brooks WM (2005) Effects of ketamine on anterior cingulate glutamate metabolism in healthy humans: a 4-T proton MRS study. Am J Psychiatry 162:394–396
Rowland LM, Beason-Held L, Tamminga CA, Holcomb HH (2010) The interactive effects of ketamine and nicotine on human cerebral blood flow. Psychopharmacology (Berl) 208:575–584
Sassano-Higgins S, Baron D, Juarez G, Esmaili N, Gold M (2016) A review of ketamine abuse and diversion. Depress Anxiety 33:718–727
Schmaal L, Hibar D, Sämann P, Hall G, Baune B, Jahanshad N, Cheung J, Van Erp T, Bos D, Ikram M (2017) Cortical abnormalities in adults and adolescents with major depression based on brain scans from 20 cohorts worldwide in the ENIGMA major depressive disorder working group. Mol Psychiatry 22:900
Seidman LJ, Valera EM, Makris N (2005) Structural brain imaging of attention-deficit/hyperactivity disorder. Biol Psychiatry 57:1263–1272

Sparks B, Friedman S, Shaw D, Aylward EH, Echelard D, Artru A, Maravilla K, Giedd J, Munson J, Dawson G (2002) Brain structural abnormalities in young children with autism spectrum disorder. Neurology 59:184–192

Stoeckel LE, Garrison KA, Ghosh SS, Wighton P, Hanlon CA, Gilman JM, Greer S, Turk-Browne NB, deBettencourt MT, Scheinost D (2014) Optimizing real time fMRI neurofeedback for therapeutic discovery and development. NeuroImage Clin 5:245–255

Strakowski SM, DelBello MP, Sax KW, Zimmerman ME, Shear PK, Hawkins JM, Larson ER (1999) Brain magnetic resonance imaging of structural abnormalities in bipolar disorder. Arch Gen Psychiatry 56:254–260

Sun L, Li Q, Li Q, Zhang Y, Liu D, Jiang H, Pan F, Yew DT (2014) Chronic ketamine exposure induces permanent impairment of brain functions in adolescent cynomolgus monkeys. Addict Biol 19:185–194

Sutherland MT, McHugh MJ, Pariyadath V, Stein EA (2012) Resting state functional connectivity in addiction: lessons learned and a road ahead. Neuroimage 62:2281–2295

Symms M, Jäger H, Schmierer K, Yousry T (2004) A review of structural magnetic resonance neuroimaging. J Neurol Neurosurg Psychiatry 75:1235–1244

Tang W, Liang H, Lau C, Tang A, Ungvari GS (2013) Relationship between cognitive impairment and depressive symptoms in current ketamine users. J Stud Alcohol Drugs 74:460–468

Tang J, Liao Y, He H, Deng Q, Zhang G, Qi C, Cui H, Jiao B, Yang M, Feng Z (2015a) Sleeping problems in Chinese illicit drug dependent subjects. BMC Psychiatry 15:28

Tang J, Morgan HL, Liao Y, Corlett PR, Wang D, Li H, Tang Y, Chen J, Liu T, Hao W (2015b) Chronic administration of ketamine mimics the perturbed sense of body ownership associated with schizophrenia. Psychopharmacology (Berl) 232:1515–1526

Tsai TH, Cha TL, Lin CM, Tsao CW, Tang SH, Chuang FP, Wu ST, Sun GH, Yu DS, Chang SY (2009) Ketamine-associated bladder dysfunction. Int J Urol 16:826–829

Valentine GW, Mason GF, Gomez R, Fasula M, Watzl J, Pittman B, Krystal JH, Sanacora G (2011) The antidepressant effect of ketamine is not associated with changes in occipital amino acid neurotransmitter content as measured by [1H]-MRS. Psychiatry Res Neuroimaging 191:122–127

Van Erp TG, Walton E, Hibar DP, Schmaal L, Jiang W, Glahn DC, Pearlson GD, Yao N, Fukunaga M, Hashimoto R (2018) Cortical brain abnormalities in 4474 individuals with schizophrenia and 5098 control subjects via the enhancing neuro imaging genetics through meta analysis (ENIGMA) consortium. Biol Psychiatry 84:644–654

Volkow ND, Wang G-J, Fowler JS, Tomasi D, Baler R (2014) Neuroimaging of addiction. In: Imaging of the human brain in health and disease. Elsevier, Amsterdam, pp 1–26

Wang C, Zheng D, Xu J, Lam W, Yew D (2013) Brain damages in ketamine addicts as revealed by magnetic resonance imaging. Front Neuroanat 7:23

Wood R, Bassett K, Foerster T, Spry C, Tong L (2012) 1.5 tesla magnetic resonance imaging scanners compared with 3.0 tesla magnetic resonance imaging scanners: systematic review of clinical effectiveness. CADTH Technol Overviews 2:e2201

Yang Y, Cui Y, Sang K, Dong Y, Ni Z, Ma S, Hu H (2018) Ketamine blocks bursting in the lateral habenula to rapidly relieve depression. Nature 554:317

Yeung L, Wai MS, Fan M, Mak Y, Lam W, Li Z, Lu G, Yew DT (2010) Hyperphosphorylated tau in the brains of mice and monkeys with long-term administration of ketamine. Toxicol Lett 193:189–193

Young LE, Bartram D, Diamond MJ, Gregg AS, Jones R (1993) Clinical evaluation of an infusion of xylazine, guaifenesin and ketamine for maintenance of anaesthesia in horses. Equine Vet J 25:115–119

Yu H, Li Q, Wang D, Shi L, Lu G, Sun L, Wang L, Zhu W, Mak YT, Wong N (2012) Mapping the central effects of chronic ketamine administration in an adolescent primate model by functional magnetic resonance imaging (fMRI). Neurotoxicology 33:70–77

Zanos P, Moaddel R, Morris PJ, Georgiou P, Fischell J, Elmer GI, Alkondon M, Yuan P, Pribut HJ, Singh NS (2016) NMDAR inhibition-independent antidepressant actions of ketamine metabolites. Nature 533:481

Zou X, Patterson TA, Sadovova N, Twaddle NC, Doerge DR, Zhang X, Fu X, Hanig JP, Paule MG, Slikker W (2009) Potential neurotoxicity of ketamine in the developing rat brain. Toxicol Sci 108:149–158

Management of Complications of Ketamine Abuse

Steffi K. K. Yuen and Anthony C. F. Ng

Abstract Since 1985, ketamine, a frequently used human and veterinary anaesthetic, has been listed as an essential medicine by the World Health Organization. This *N*-methyl-D-aspartate receptor antagonist has also been used and studied extensively for its role in the treatment of depression. Owing to its potential implications on health when used illicitly, at least 60 countries have already put ketamine under national control. Its illicit use has been reported on a global scale, so are its delirious complications from chronic abuse. The detrimental effects of ketamine abuse encompass different organ systems, where long-lasting complications on the genitourinary, gastrointestinal, hepatobiliary and neuropsychiatric systems stood out. In this relatively new entity of ketamine-associated complications, the importance of proper assessment and building of good rapport is emphasized. Abstinence is the first key for the successful management of its complications. Both medical and surgical management have its role in the multidisciplinary management.

Keywords Ketamine · Uropathy · Complications · Cystitis · Management

The detrimental effects of ketamine misuse was only evident to the medical world within the recent decade or so. Despite its initial formal reporting of toxic complications on genitourinary system by Shahani et al. (2007) from Canada and Chu et al. (2008) from Hong Kong, the ketamine abuse problem has gravitated farther and was popularized in Asian countries, particularly in Hong Kong SAR. Ketamine, an *N*-methyl-D-aspartate receptor antagonist, whilst being utilized clinically as an anaesthetic agent, is also being commonly misused as a recreational drug since the late 1990s.

The detrimental effects of ketamine abuse encompass many organ systems, where long-lasting complications on genitourinary, gastrointestinal, hepatobiliary and neuropsychiatric systems stood out. Its alarming burden on health care,

---

S. K. K. Yuen · A. C. F. Ng (✉)
Department of Surgery, SH Ho Urology Centre, The Chinese University of Hong Kong, Hong Kong, China
e-mail: ngcf@surgery.cuhk.edu.hk

K. Hashimoto et al. (eds.), *Ketamine*,
https://doi.org/10.1007/978-981-15-2902-3_3

economic and societal fronts has warranted measures to help cracking down the problem.

1 Why Is Managing Complications an Important Topic?

Ketamine-associated complications revealed as a relatively new entity to the medical world. It primarily affects young to middle-aged ketamine abusers, amongst whom they popularized ketamine as a "post-clubbing drug" where you can "sit down and float". Its handy availability, easy administration and psycho-stimulant properties render it attractive as a recreational drug. The exact pathophysiology and effects of ketamine and its metabolites are yet to be completely dissected and laid out. This poses as both a medical and a social issue to be tackled.

More than 50 countries have reported the presence of ketamine in illicit drug markets from the annual United Nation Office on Drug and Crime World Drug Reports. The 2017/2018 Crime Survey for English and Wales reported a twofold increase in illicit ketamine usage from 0.4 to 0.8% amongst adults aged 16–59 years, equating to 141,000 more using the drug than previous year (Home Office (U.K.) NS 2018). Such estimate of ketamine usage is the highest since survey first began in 2006/07 for this drug. The burden of ketamine abuse lies heavily in the group aged 16–24 years, citing a rise from 1.2 to 3.1%. In Hong Kong, ketamine was the most commonly abused psychotropic drug during 2008–2014. Methamphetamine surpassed ketamine to become the most popular psychotropic substance abused since 2015, followed by triazolam, midazolam, zopiclone and cocaine in 2017 (Narcotics Division SBTGotHKSAR 2019). Over the years, the concerted efforts of governmental officials have helped brought the numbers down by dismantling clandestine ketamine laboratories and decreasing their supply sources, mainly in East and South-East Asia (e.g. China and Malaysia).

Although the number of reported abusers of psychotropic substances is on the fall, the complications are still slowly surfacing. Amongst such young population of ketamine abusers, untreated or suboptimally treated complications will serve as returning patients when their health and quality of life are detrimentally affected.

A vast majority of ketamine abusers are disguised amongst the community snorting the drug undercover for years unnoticed by closed ones. Only by concerted efforts from a dedicated team of clinicians, paediatricians, psychiatrists, psychologists, social workers, teachers, nurses and therapists can we truly evaluate and manage the full spectrum of ketamine-associated complications. Family and parental support, particularly for the teenage abusers, is of paramount importance to keep the ketamine abuser on track, abstaining from the drug and compliant to follow up and treatments. Complications involving genitourinary, hepatobiliary, gastrointestinal and neuropsychiatric systems mark ketamine-associated complications as a complex class of disease entity.

2 Pathophysiology

Ketamine is converted to active metabolic norketamine via hepatic biotransformation through the cytochrome P450 and then eliminated via the hepatic route into conjugated hydroxyl metabolites, which is then excreted renally (Dinis-Oliveira 2017). The damage of ketamine and its metabolites on the genitourinary tract is proposed as via direct toxic damage, autoimmune reactions and/or microvascular reactions.

Inflammation of bladder epithelium, denuding of urothelium neovascularization and petechial haemorrhage of the bladder are all reported. However, the exact pathophysiology is not fully known. Histologically, four out of nine Canadian patients in the first series had biopsies revealing denuding of urothelium and inflammation of bladder epithelium with a mild eosinophilic infiltrate (Shahani et al. 2007). The histopathology of bladder biopsies from ketamine cystitis patients shows features of chronic inflammation with strong resemblance to those in interstitial cystitis. The predominant type of infiltrating inflammatory cells is lymphocytes. The severity of histopathology of inflammatory cell infiltration of bladder mucosa is shown to be associated with clinical symptomatology. Patients with moderate or severe neutrophils or lymphocytes infiltration in bladder mucosa had significantly more severe bladder pain and smaller bladder capacity (Jhang et al. 2018). In rat ketamine cystitis models, it is shown that increasing fibrosis and submucosal apoptosis were found according to escalating dose of ketamine. It is postulated that bladder fibrotic change contributes to the manifestation of lower urinary tract symptoms (Song et al. 2016).

3 Complications Related to Urinary System: Ketamine-Associated Uropathy

Ketamine abuse may affect both the upper and lower urinary tract. In fact, its effects may go unnoticed until late in stage where irreversible damages have already taken place. Lower urinary tract symptoms, cystitis, bladder dysfunction, contracted bladder, ureteric strictures and secondary renal damage have all been reported in chronic ketamine abusers. The clinical syndrome of ketamine-associated uropathy is described as a small, painful bladder, associated with incontinence, upper tract obstruction with or without papillary necrosis (Shahani et al. 2007; Chu et al. 2008).

4 Symptomatology

In the largest online survey to date, more than a quarter of ketamine abusers reported experiencing urinary symptoms to variable extent (Winstock et al. 2012). The onset of symptoms varies from few days to years. From one of the largest reported series, the mean duration of ketamine usage was 81 months prior to the presentation of voiding symptoms (Tam et al. 2014). Ketamine-associated lower urinary tract symptom is recognized as ketamine-induced cystitis. Painful frequent small volume void is a classic chief complaint. Irritative symptoms, such as frequency, urgency, nocturia, dysuria, urge incontinence, pelvic pain and painful hematuria, vary in severity. Symptomatology could be very debilitating to a young individual. Urinary frequency could be down to every quarter of an hour, barring the individual from normal social activities. In more severe cases, the young individuals could be incontinent and dependent on diapers to avoid interruptions to the washroom during work hours. In a 463 patient cohort looking into both active and inactive ketamine abusers by Yee et al. (2015), it was shown that active abusers have significant higher pelvic pain and frequency (PUF) score than inactive ones (23.3 ± 6.7 vs. 19.8 ± 7.7; $p < 0.0005$). Cystitis and bladder dysfunction are the common manifestations. However, its effects are observed to involve organs beyond the bladder.

Upper tract pathologies may be asymptomatic in many ketamine abusers, or may manifest most commonly as loin discomfort. In a cohort of 572 ketamine-associated uropathy patients, up to 16.8% of the patients were found to have unilateral or bilateral hydronephrosis on ultrasonography (Yee et al. 2017), secondary to inflammatory reaction, stricture, vesicoureteric reflux or papillary necrosis (Fig. 1). Functional bladder capacity (OR 0.997, $p = 0.029$), serum creatinine > 100 μmol/L (OR 1.238, $p = 0.016$) and a deranged serum liver enzyme profile (OR 1.967, $p = 0.006$) are predictive factors of hydronephrosis in ketamine abusers from a multi-variate analysis. Obstructive uropathy, renal impairment and end-stage renal disease are uncommon sequalae of ketamine-associated uropathy.

Symptoms are not confined urologically; they extend to the domains of sexual dysfunction in both female and male patients (Jang et al. 2012; Yang et al. 2018). In a case-control study of female ketamine abusers by Jang et al. (2012), with the exception of the sexual desire domain in female patients, the abusers scored lower on the arousal, lubrication, orgasm, satisfaction and pain domains of the Female Sexual Function Index score (17.65 ± 6.15 vs. 25.87 ± 4.16 for controls ($p < 0.001$). For the male counterparts, Yang et al. (2018) reported amongst 1056 abusers (993 street ketamine abusers who presented to the urology and 63 who presented to the hospital), erectile dysfunction (30.8%) is frequently observed with International Index of Erectile Function (IIEF-5) score ≤21. Multi-variate analysis revealed age is ≥30 years old (OR = 1.765) as a risk factor for male erectile dysfunction; subgroup analysis revealed abstinence for 3 months or more as a protective factor. A small study evaluated the effects of ketamine on membrane integrity, DNA fragmentation and sperm parameters in humans (Absalan et al. 2014). It is shown to have significantly lower total sperm motility, decreased sperm viability and

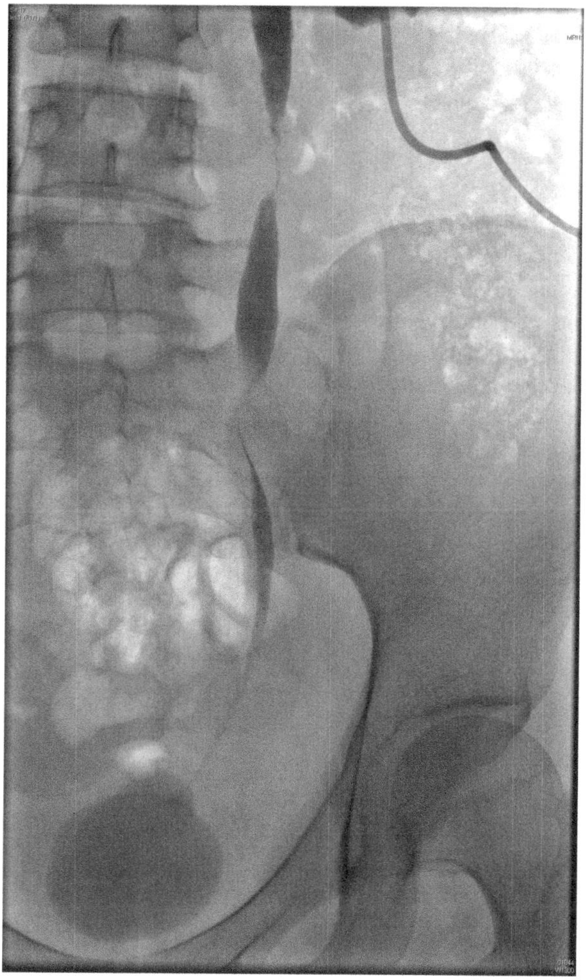

Fig. 1 shows multiple left ureteric strictures and small contracted bladder on antegrade pyelogram

abnormal sperm parameters in progressive motility. Since this is a relatively young patient group where individuals may have not settled down, families may not be completed. Whether infertility will pose an important issue remains a question to be answered. Both urinary symptoms and sexual dysfunction certainly are shown to have adversely impacted on this group of patient's quality of life.

5 Assessment, Diagnosis and Investigation

Diagnosis is often easily made clinically.

5.1 *History Taking*

History taking is an important step that should not be underestimated in the management of ketamine abusers. It helps develop doctor–patient rapport, draws patient's compliance and often reveals the fundamental societal or personal psychological reasons that steer the individual down the illicit substance pathway. A detailed history taking of symptomatology, duration and frequency of recreational ketamine usage, social support and background is an important step of assessment. Of note during history taking, the reported consumption frequency by patients may not be as reliable as the monetary value spent on street illicit ketamine purchase. Misuse of ketamine >3 times per week is associated with significantly lower voided volumes. Pelvic pain, frequency and urgency are reported to be significantly higher in those with chronic ketamine abuse >24 months, compared to those with shorter durations (Mak et al. 2011).

5.2 *Initial Assessment and Workup*

In the largest prospective cohort of ketamine-associated uropathy, Yee et al. (2015) proposed a standardized management protocol at a dedicated urological clinic. Invasive investigations too early on in the assessment process may deter this group of patients from adhering to the programme and follow-up schedules.

5.2.1 Standardized Questionnaire

Apart from the commonly used International Prostate Symptom Score (IPSS), IIEF-5 and frequency/volume charts, symptom assessment can be objectively performed by means of pelvic pain and urgency or frequency (PUF) symptom scale, the EuroQol visual analog scale.

Pelvic pain and urgency/frequency (PUF) symptom scale, a tool initially used in the assessment of interstitial cystitis, correlates with worse symptomatology (Yee et al. 2015). The questionnaire comprises seven questions: daytime and night-time frequency, whether the pain affects or is present during sexual intercourse, pelvic pain and its severity, urgency and its degree of severity. It includes a symptom score and a bother score, totalling 35 points maximum. Its Chinese version has been validated and used in the assessment of patients with ketamine abuse (Ng et al. 2012).

This entity has strong resemblance between interstitial cystitis and ketamine-associated cystitis, hence PUF symptom scale has demonstrated its role in the assessment of such patient groups.

The EuroQol visual analog scale (EQ VAS) is a visual scale marked 0–100 for patients' subjective assessment of own health state. The lower the score, the worse the patients self-perceive their own health state. Such standardized questionnaires are important in the assessment and follow-up of patient's progress.

5.2.2 Initial Workup

Initial workup should be kept non-invasive, including uroflowmetry, urine microscopy and culture and serum creatinine, which will help to build the rapport and trust between the patients and the health care providers. As many patients already suffered from painful urination, invasive investigation might frighten them and lead to poor compliance and also delay in treatment. Functional bladder capacity can be calculated by the summation of the voided volume and post-void residual urine volume during uroflowmetry assessment. In a cross-sectional prospective cohort, the mean voided volume of ketamine abusers is reported up to 111.5 mL, with a mean bladder capacity of 152.5 mL (Tam et al. 2014).

Ultrasonography of the urinary system can be performed to screen for any sign of obstructive uropathy. Hydronephrosis can be found in 8.1–51% (Chu et al. 2008; Yee et al. 2017; Tam et al. 2014) amongst which 10.4–30.7% may have concomitant renal impairment. Small bladder volume and wall thickening are also common features (Mason et al. 2010).

Computer tomography with urogram phase is also very useful because upper tract involvement is not uncommon in this group. The common CT findings include diffuse bladder wall thickening, small bladder volume and perivesical inflammation. Moreover, they can also pick up upper tract involvement, such as unilateral/bilateral hydronephrosis and ureteric wall thickening (Huang et al. 2014).

Cystoscopy is invasive and not essential in the diagnosis and initial management of ketamine cystitis, as the diagnosis is based mostly on clinical condition. Unless there are clinical suspicious of other pathology, or failed response to initial therapy, endoscopy might be considered. Endoscopy findings vary from normal-looking bladder mucosa to contracted bladder with erythematous cystitis (Tam et al. 2014), ulceration, neovascularization and petechial haemorrhage (Chu et al. 2008).

Video-urodynamic study can be reserved for patients who require in-depth evaluation of bladder condition and detection of vesicoureteric reflux. It is a particularly useful tool as part of the preoperative planning when all conservative and medical measures fail. Typical findings include diminished bladder compliance or presence of detrusor overactivity with or without urinary leakage when the bladder is filled to a small capacity of 30–50 mL (Chu et al. 2008) (Fig. 2).

With appropriate investigations, prompt detection and management of ketamine-associated uropathy can be carried out.

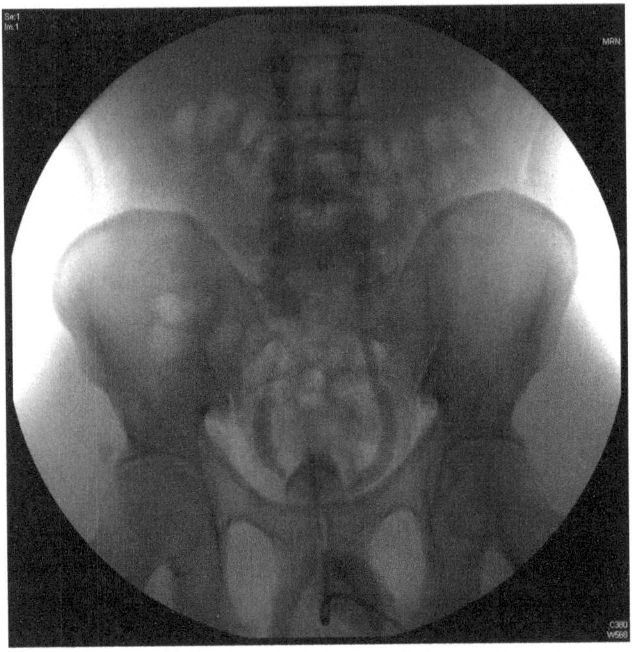

Fig. 2 shows small contracted bladder with size of a Foley's catheter balloon and bilateral vesico-ureteric reflux on cystogram

Investigations

- Uroflowmetry, functional bladder capacity
- Urine microscopy and culture
- Serum creatinine
- Liver function test
- Ultrasonography of urinary system
- Computer tomography with urogram phase
- Cystoscopy
- Video-urodynamic study

5.3 Management

Multidisciplinary approach is of paramount importance in successfully treating both the clinical and psychological aspects of the disease entity.

5.3.1 Abstinence of Ketamine

First and foremost, abstinence of ketamine is emphasized to patients during their first visit to health care. There is a dose and frequency response relationship between ketamine use and urinary symptoms. Symptomatology score is improved with a direct positive relationship with the length of abstinence from the drug. There is potential to normalize functional damages beyond 1 year of ketamine cessation. More than half reported improvement of urinary symptoms upon abstaining (Winstock et al. 2012). Those abstained have a lower symptom score (19.3 vs. 24.1; $p < 0.001$), a larger voided volume (126 vs. 85 mL; $p < 0.001$) and a larger bladder capacity (204.8 vs. 126.7 mL; $P < 0.001$) compared with active abusers (Tam et al. 2014). Reduced benefits from ketamine abstinence is observed if the drug was misused at higher frequencies or longer duration.

Abstinence is easier said than done. Working hand in hand with a dedicated team of social worker, psychologist, psychiatrist, teacher, parents and family could result in higher rates of success. Local support from drug and addiction services can also play a decisive role in abstinence and treatment success.

5.3.2 Treatment Ladder

The four-tier treatment protocol has been developed and implemented at the dedicated Youth Urological Treatment Centre reported by Yee et al. (2015), in hope of offering standardized treatment multidisciplinary care to these batch of patients (Hong et al. 2018). The stepwise approach proposed is as follows:

Tier one: NSAIDS/ COX-2 inhibitors/anti-cholinergics
Tier two: Pregabalin or a short course of opioid analgesics
Tier three: A course of intravesical instillation of sodium hyaluronate
Tier four: Surgical intervention

5.3.3 Medical Treatment

Symptoms are often refractory to treatment with antibiotics, simple analgesics and anti-cholinergics alone. First-line oral medications include nonsteroidal anti-inflammatory drugs (e.g. diclofenac and etoricoxib) and anticholinergic agents (e.g. solifenacin). Beta-3 adrenoceptor agonists can be another option.

Phenazopyridine and paracetamol are used for pain control. The aim is to cover the possible increase in pain experienced during the initial abstinence period. If first-line treatment fails to provide sufficient analgesic relief, the second-line treatment is introduced and added onto the cocktail of medications – pregabalin or a short course of opioid group of analgesics (e.g. tramadol). Regular follow-up and reassessments are carried out to ensure abstinence and adherence to the treatment regime. Outcome is assessed with functional bladder capacity, pelvic pain and urgency or frequency (PUF) symptom scale and the EuroQol visual analog scale.

Both abstinence from ketamine usage and the amount of ketamine consumed are factors predicting the improvement of PUF scores. For patients who required second-line oral therapy in that cohort, 67.7% reported improvement in symptoms.

5.3.4 Intravesical Instillation of Sodium Hyaluronate

There have been case reports with complete resolution of symptoms, as well as significant improvement in voided volume for the patients after intravesical treatment with sodium hyaluronic acid.

5.3.5 Surgical Treatment

Urinary tract reconstruction has proved a surgical challenge for ketamine-associated uropathy. Even when managed in tertiary high-volume reconstructive units, this group of patients are still at high risk of significant perioperative complications (Sihra et al. 2018). Meticulous preoperative assessment and multidisciplinary approach to optimize treatment strategies are recommended.

Surgical correction is reserved for patients with confirmed abstinence from ketamine use. Whenever indicated, nephrostomy or ureteric stents are inserted to preserve renal function until definitive surgical correction is allowed. There is no optimal time recommended for abstinence from ketamine abuse before major reconstructive use can be carried out.

Common indications for surgical reconstruction include:

- Small contracted bladder
- Medically uncontrolled pelvic pain or lower urinary tract symptoms
- High-pressure, low-bladder compliance

Reported options for surgical intervention:

- Intra-detrusor botulinum toxin injection
- Hydrodistension
- Augmentation cystoplasty with or without Mitrofanoff channels
- Ileal conduit urinary diversion
- Cystectomy with neobladder

Ureteric stricture

- Metallic stents
- Ureteric dilatation
- Ureteric reimplantation
- Autotransplantation

There are reports where intra-detrusor onabotulinum toxin injection was ineffective in symptom relief. Because ketamine cystitis shares similar features with interstitial cystitis, hydrodistension was carried out in small case reports with unsatisfactory outcomes.

Surgical correction was the most commonly reported in the literature with evident outcomes. The number of reported surgical corrections is small, where complication rates are high, even in high-volume tertiary centres. A retrospective cohort where 44 patients spanning over a decade were reviewed at a high-volume tertiary reconstructive unit, 14 patients underwent major reconstruction with indications including intractable symptoms, high-pressure compliance loss with upper tract damage and ureteric obstruction (Sihra et al. 2018). Surgical intervention included ileal conduit urinary diversion, augmentation cystoplasty with or without Mitrofanoff channels, ureteric re-implantation and cystectomy with neobladders. Complications included anastomotic leaks, ureteric strictures, adhesive small bowel obstruction, renal failure and sepsis.

Another cohort of patients where augmentation cystoplasty was performed was reported by Ng et al. (2013), where it reached the same conclusion on high complication rates and the high tendency of resuming ketamine abuse after surgery. Although augmentation cystoplasty is a very effective way of increasing bladder capacity and relieving storage lower urinary tract symptoms (Figs. 3 and 4), it was discussed that the option of simple non-continent urinary diversion (e.g. ileal conduit) may be a better option as it has faster recovery. However, it may not be acceptable and appealing to this young population of patients. The extremely high

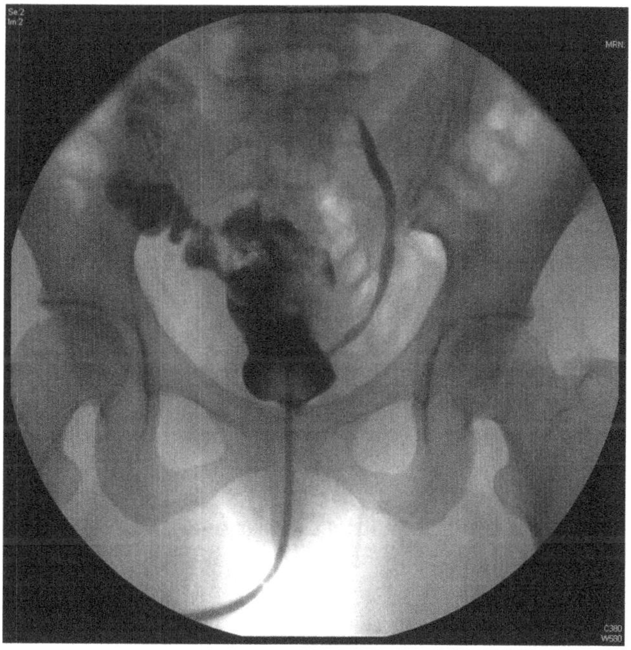

Fig. 3 shows an early post-operative cystogram of a patient who received ileal interposition for right ureteric stricture and augmentation ileocystoplasty

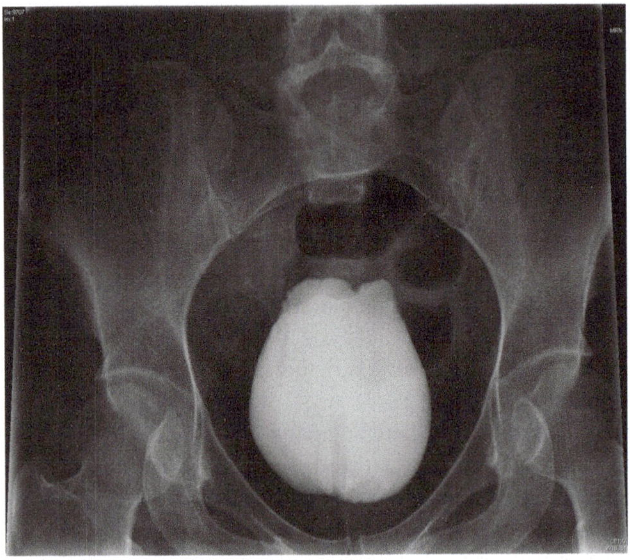

Fig. 4 is a cystogram revealing a patient who have received augmentation ileocystoplasty

rates of indulging back into ketamine abuse should caution urologists offering surgical definitive treatment in discretely selective groups that have abstained from ketamine abuse and have social support in continuing the abstinence so that all efforts will not go waste.

Ketamine-associated uropathy is an evolving disease entity, patient centred tailored management will provide the best outcomes.

6 Complications Involving the Gastrointestinal and Hepatobiliary Systems

There are also reports on gastrointestinal changes in ketamine abusers including epigastric pain, hepatic dysfunction and impaired gallbladder activity. Many ketamine abusers are often disguised amongst the community. They may not present to clinicians until unbearable symptoms or complications arise. Clinicians may seize the opportunity to identify disguised abusers when being consulted for non-specific symptoms such as epigastric discomfort and lower urinary tract symptoms.

Up to a quarter, amongst 611 patients who have sought medical consultation for ketamine-associated uropathy in a cross-sectional study by Liu et al. (2017), reported upper gastrointestinal symptoms. There may be up to 5 years lag from ketamine abuse to symptoms' onset. Symptomatology may vary from epigastric pain (25.4%), recurrent vomiting (7.9%), anaemia (5.9%) and gastrointestinal bleeding (3.3%). As upper gastrointestinal symptoms usually precede the presentation of voiding symptom, direct enquiry on substance usage in young non-helicobacter gastritis or ulcer might help to identify potential hidden abuser and allow earlier intervention to them.

Distinct biliary anomalies on magnetic resonance cholangiography patterns have been reported (Seto et al. 2018). Magnetic resonance cholangiography showed biliary tract anomalies in up to 61.9%, where an elevated alkaline phosphatase level was observed. Three distinct radiological patterns were reported: (1) diffuse dilatation of extrahepatic ducts; (2) fusiform dilatation of extra-hepatic ducts with distal tapering and (3) intrahepatic duct dilatation or bleeding and strictures with normal extrahepatic ducts. Elevated alkaline phosphatase level and lack of concomitant drug are predictive of biliary anomalies. Such biliary anomalies are reversible after ketamine abstinence. Biliary sepsis and decompensated cirrhosis are rare sequalae.

Recreational ketamine inhalation serves as a dualistic challenge on health care and social fronts worldwide. Early recognition and detection of ketamine-associated uropathy allows timely management. A combined approach of medical treatment and psychosocial support is of paramount influence in successful abstinence. Surgical intervention is only indicated when abstinence is confirmed and deemed appropriate timing.

References

Absalan F, Ghannadi A, Zabihi A (2014) The effects of different doses of ketamine on quality of normal ejaculated sperm. Int J Fertil Steril 8(2):207–214

Chu PS, Ma WK, Wong SC, Chu RW, Cheng CH, Wong S, Tse JM, Lau FL, Yiu MK, Man CW (2008) The destruction of the lower urinary tract by ketamine abuse: a new syndrome? BJU Int 102(11):1616–1622. https://doi.org/10.1111/j.1464-410X.2008.07920.x

Dinis-Oliveira RJ (2017) Metabolism and metabolomics of ketamine: a toxicological approach. Forensic Sci Res 2(1):2–10. https://doi.org/10.1080/20961790.2017.1285219

Home Office (U.K.) NS (2018) Drug misuse: findings from the 2017/18 Crime Survey for England and Wales. https://assets.publishing.service.gov.uk/government/uploads/system/uploads/attachment_data/file/729249/drug-misuse-2018-hosb1418.pdf. Accessed 11 Jun 2019

Hong YL, Yee CH, Tam YH, Wong JH, Lai PT, Ng CF (2018) Management of complications of ketamine abuse: 10 years' experience in Hong Kong. Hong Kong Med J 24(2):175–181. https://doi.org/10.12809/hkmj177086

Huang LK, Wang JH, Shen SH, Lin AT, Chang CY (2014) Evaluation of the extent of ketamine-induced uropathy: the role of CT urography. Postgrad Med J 90(1062):185–190. https://doi.org/10.1136/postgradmedj-2013-131776

Jang MY, Long CY, Chuang SM, Huang CH, Lin HY, Wu WJ, Juan YS (2012) Sexual dysfunction in women with ketamine cystitis: a case-control study. BJU Int 110(3):427–431. https://doi.org/10.1111/j.1464-410X.2011.10780.x

Jhang JF, Hsu YH, Jiang YH, Lee CL, Kuo HC (2018) Histopathological characteristics of ketamine-associated uropathy and their clinical association. NeurourolUrodyn 37(5):1764–1772. https://doi.org/10.1002/nau.23514

Liu SYW, Ng SKK, Tam YH, Yee SCH, Lai FPT, Hong CYL, Chiu PWY, Ng EKW, Ng CF (2017) Clinical pattern and prevalence of upper gastrointestinal toxicity in patients abusing ketamine. J Dig Dis 18(9):504–510. https://doi.org/10.1111/1751-2980.12512

Mak SK, Chan MT, Bower WF, Yip SK, Hou SS, Wu BB, Man CY (2011) Lower urinary tract changes in young adults using ketamine. J Urol 186(2):610–614. https://doi.org/10.1016/j.juro.2011.03.108
Mason K, Cottrell AM, Corrigan AG, Gillatt DA, Mitchelmore AE (2010) Ketamine-associated lower urinary tract destruction: a new radiological challenge. Clin Radiol 65(10):795–800. https://doi.org/10.1016/j.crad.2010.05.003
Narcotics Division SBTGotHKSAR (2019) Central Registry of drug abuse sixty-seventh report 2008-2017 (Chapter 2). https://www.nd.gov.hk/pdf/report/crda_67th/Chapter2%20Trends%20(67th).pdf. Accessed 20 May 2019
Ng CM, Ma WK, To KC, Yiu MK (2012) The Chinese version of the pelvic pain and urgency/frequency symptom scale: a useful assessment tool for street-ketamine abusers with lower urinary tract symptoms. Hong Kong Med J 18(2):123–130
Ng CF, Chiu PK, Li ML, Man CW, Hou SS, Chan ES, Chu PS (2013) Clinical outcomes of augmentation cystoplasty in patients suffering from ketamine-related bladder contractures. Int Urol Nephrol 45(5):1245–1251. https://doi.org/10.1007/s11255-013-0501-4
Seto WK, Mak SK, Chiu K, Vardhanabhuti V, Wong HF, Leong HT, Lee PSF, Ho YC, Lee CK, Cheung KS, Yuen MF, Leung WK (2018) Magnetic resonance cholangiogram patterns and clinical profiles of ketamine-related cholangiopathy in drug users. J Hepatol 69(1):121–128. https://doi.org/10.1016/j.jhep.2018.03.006
Shahani R, Streutker C, Dickson B, Stewart RJ (2007) Ketamine-associated ulcerative cystitis: a new clinical entity. Urology 69(5):810–812. https://doi.org/10.1016/j.urology.2007.01.038
Sihra N, Ockrim J, Wood D (2018) The effects of recreational ketamine cystitis on urinary tract reconstruction—a surgical challenge. BJU Int 121(3):458–465. https://doi.org/10.1111/bju.14094
Song M, Yu HY, Chun JY, Shin DM, Song SH, Choo MS, Song YS (2016) The fibrosis of ketamine, a noncompetitive N-methyl-D-aspartic acid receptor antagonist dose-dependent change in a ketamine-induced cystitis rat model. Drug Chem Toxicol 39(2):206–212. https://doi.org/10.3109/01480545.2015.1079916
Tam YH, Ng CF, Pang KK, Yee CH, Chu WC, Leung VY, Wong GL, Wong VW, Chan HL, Lai PB (2014) One-stop clinic for ketamine-associated uropathy: report on service delivery model, patients' characteristics and non-invasive investigations at baseline by a cross-sectional study in a prospective cohort of 318 teenagers and young adults. BJU Int 114(5):754–760. https://doi.org/10.1111/bju.12675
Winstock AR, Mitcheson L, Gillatt DA, Cottrell AM (2012) The prevalence and natural history of urinary symptoms among recreational ketamine users. BJU Int 110(11):1762–1766. https://doi.org/10.1111/j.1464-410X.2012.11028.x
Yang SS, Jang MY, Lee KH, Hsu WT, Chen YC, Chen WS, Chang SJ (2018) Sexual and bladder dysfunction in male ketamine abusers: a large-scale questionnaire study. PLoS One 13(11):e0207927. https://doi.org/10.1371/journal.pone.0207927
Yee CH, Lai PT, Lee WM, Tam YH, Ng CF (2015) Clinical outcome of a prospective case series of patients with ketamine cystitis who underwent standardized treatment protocol. Urology 86(2):236–243. https://doi.org/10.1016/j.urology.2015.05.003
Yee CH, Teoh JY, Lai PT, Leung VY, Chu WC, Lee WM, Tam YH, Ng CF (2017) The risk of upper urinary tract involvement in patients with ketamine-associated uropathy. Int Neurourol J 21(2):128–132. https://doi.org/10.5213/inj.1732704.352

Site of Ketamine Action on the NMDA Receptor

Hisashi Mori

Abstract Ketamine, an antagonist of *N*-methyl-D-aspartate receptors (NMDARs), produces rapid and sustained reduction of symptoms in patients with treatment-resistant depression. NMDARs are critical for neural network formation, neuronal plasticity, higher brain functions, and pathophysiology of neurodegenerative and psychiatric disorders. Recent studies have identified functional domains of diverse NMDAR subunits, as well as the site of ketamine action on NMDARs. The site of ketamine action overlaps with the site of physiological voltage-dependent Mg^{2+} block. Furthermore, different NMDAR GluN2 subunits contribute differentially to the sensitivity of ketamine. High-resolution analyses of the structure of the action site of ketamine on NMDARs and the mechanisms of ketamine action in vivo will contribute to the development of novel and effective antidepressant drugs.

Keywords Ketamine · Depression · Antidepressant · *N*-methyl-D-aspartate receptors · GluN1 · GluN2 · GluN3 · Glutamate · D-serine · Glycine · Gene knockout mice

Abbreviations

AMPA	α-Amino-3-hydroxy-5-methyl-4-isoxazole propionic acid
APV	D-2-Amino-5-phosphono-valerate
ATD	Amino-terminal domain
CNS	Central nervous system
CTD	Carboxy-terminal domain
GluR	Glutamate receptor
KO	Gene knockout
LBD	Ligand-binding domain
LTP	Long-term potentiation
NMDA	*N*-methyl-D-aspartate
PCP	Phencyclidine

H. Mori (✉)
Department of Molecular Neuroscience, Graduate School of Medicine and Pharmaceutical Sciences, University of Toyama, Toyama, Japan
e-mail: hmori@med.u-toyama.ac.jp

K. Hashimoto et al. (eds.), *Ketamine*,
https://doi.org/10.1007/978-981-15-2902-3_4

PFC Prefrontal cortex
TMD Transmembrane domain

1 Introduction

The World Health Organization provides the following overview of depression: *Depression is a common illness worldwide, with more than 300 million people affected. Depression is different from usual mood fluctuations and short-lived emotional responses to challenges in everyday life. Especially when long-lasting and with moderate or severe intensity, depression may become a serious health condition. It can cause the affected person to suffer greatly and function poorly at work, at school and in the family. At its worst, depression can lead to suicide. Close to 800,000 people die due to suicide every year.* (https://www.who.int/en/news-room/fact-sheets/detail/depression).

The depressive state is observed in bipolar (mania and depression) and major depressive disorders, and its underlying causes are the subject of ongoing intensive research. The monoaminergic hypothesis of depression predicts that depression is rooted in decreased levels of serotonin, norepinephrine, and/or dopamine in the central nervous system (CNS). Some antidepressants that target monoaminergic transmission, such as selective serotonin reuptake inhibitors (e.g., fluoxetine, sertraline, and paroxetine), induce rapid biochemical effects but require several weeks to exert clinical effects (Rush et al. 2006). In contrast, ketamine, an *N*-methyl-D-aspartate (NMDA) receptor (NMDAR) antagonist, produces rapid (within hours) and sustained reduction in depressive symptoms in patients with treatment-resistant depression (Berman et al. 2000; Zarate et al. 2006). The focus of this chapter is the site of ketamine action on NMDARs from the view of molecular biology.

2 Ketamine

Ketamine, 2-(*O*-chlorophenyl)-2-methylamino cyclohexanone hydrochloride, is a structural analog of phencyclidine (PCP), which was originally developed as a general anesthetic. Although PCP can be used as a safe and reliable anesthetic without respiratory suppression, some patients experience severe and prolonged postsurgery delirium. Thus, the use of PCP has been largely avoided in human surgery (Graifenstein et al. 1958; Johnstone et al. 1959). Ketamine was developed as a result of efforts to synthesize new, shorter-acting general anesthetics with fewer side effects (including delirium). Ketamine has about one-tenth the potency of PCP, with minimal side effects and no occurrence of severe delirium. Thus, ketamine is safer than PCP and preferred for clinical use. Ketalar (ketamine hydrochloride) was first approved by the Food and Drug Administration for human use in 1970. In human studies, high doses of ketamine (1–2 mg/kg, I.V. (Domino et al. 1984; De Simoni et al. 2013)) induced general suppression of the CNS and produced general anesthesia. In contrast, subanesthetic doses of ketamine (0.5 mg/kg I.V. over 40 min (Domino et al. 1984; De Simoni et al. 2013)) induced psychotomimetic effects. Low doses of ketamine induced a dissociative state, in which patients who receive nociceptive stimuli do not experience pain (Garfield et al. 1972; Kohrs and Durieux 1998; Bergman 1999). Auditory and visual hallucinations are common with ketamine treatment in humans; thus, ketamine was used to develop a pharmacological model of schizophrenia in animals (Olney et al. 1999; Kehrer et al. 2008). Although the causes of schizophrenia are largely unknown, the glutamate hypothesis posits that the underlying mechanisms are linked to glutamatergic signaling via NMDARs (Zhou and Sheng 2013). Furthermore, lower doses of ketamine have stimulant effects and produce mild dissociation with hallucinations and a distortion of time and space. Thus, ketamine is widely used a recreational, and often abused, drug (Morgan et al. 2009). In contrast, low doses of ketamine treatment can provide relief from chronic bipolar disorders and chronic depression (Berman et al. 2000; Zarate et al. 2006). Thus, ketamine is expected as a novel drug against these psychiatric disorders. Ketamine is a mixture of (*R*)- and (*S*)-enantiomers (Fig. 1). To understand the molecular mechanisms underlying the effects of ketamine, the critical finding that ketamine is a selective antagonist of the NMDA-type glutamate receptors (GluRs) was reported in 1983 (Anis et al. 1983).

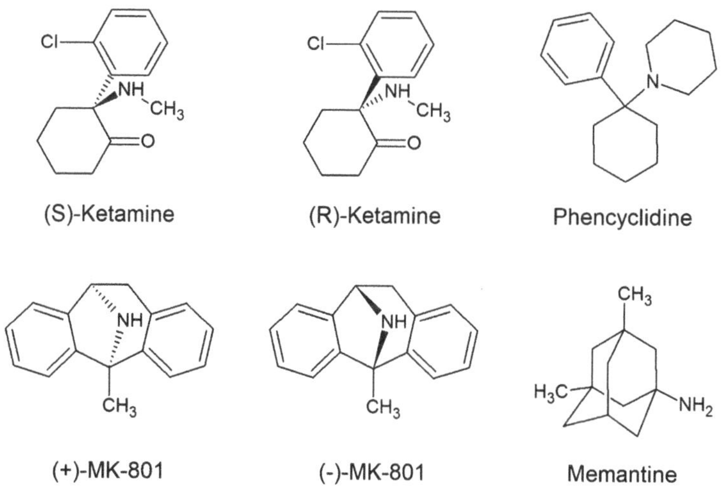

Fig. 1 Structure of open channel blockers of NMDARs. Structures of ketamine enantiomers (S-ketamine and R-ketamine), phencyclidine (PCP), MK-801 (+ and – forms), and memantine are shown

3 The NMDA Receptor

The NMDA-type GluRs are members of the glutamate-gated ion channels. Glutamate is a major excitatory neurotransmitter in the mammalian CNS and binds to receptors called GluRs. The binding of glutamate to its receptor induces excitatory neurotransmission and intracellular signal transduction. GluRs are classified as either ionotropic (iGluR) or metabotropic, based on their speed of neurotransmission or signaling mechanisms, respectively. iGluRs are further pharmacologically classified into three major subtypes: α-amino-3-hydroxy-5-methyl-4-isoxazole propionic acid (AMPA), kainate, and NMDA receptors.

NMDARs were first isolated pharmacologically using the highly selective agonist NMDA and the competitive antagonist D-2-amino-5-phosphono-valerate (APV), synthesized by Curtis and Watkins (1963). These specific ligands were used to identify the function of NMDARs and to discriminate between the roles of NMDA- and non-NMDA-type GluRs in the CNS, both in vitro and in vivo. The activation of NMDARs requires binding of an agonist (i.e., glutamate) and a co-agonist (glycine or D-serine) and release from voltage-dependent Mg^{2+} blocking. Using APV and hippocampal brain slices, Collingridge et al. (1983) identified a critical role for NMDARs in the induction of synaptic plasticity (long-term potentiation, LTP), which is thought to be a cellular mechanism of learning and memory. Furthermore, Morris et al. (1986) showed an impairment of spatial learning in rats after treatment of the hippocampus with APV. Attenuation of LTP and learning following treatment with APV indicated a critical role for NMDARs in learning and memory. Furthermore, many physiological and pathological roles of NMDARs have been examined using selective agonists and antagonists. Because NMDARs are ion channels, many open channel blockers (uncompetitive antagonists) of NMDARs have been identified, including ketamine, PCP, MK-801, and memantine (Fig. 1). In 1991, the first NMDAR subunit was identified by molecular biological approaches (Moriyoshi et al. 1991).

3.1 Molecular Diversity of NMDARs

The NMDAR subunit names used here are based on nomenclature from the International Union of Basic and Clinical Pharmacology, namely, GluN1, GluN2A, GluN2B, GluN2C, GluN2D, GluN3A, and GluN3B. These subunits are also named on the basis of genetic nomenclature as GRIN1, GRIN2A, GRIN2B, GRIN2C, GRIN2D, GRIN3A, and GRIN3B.

The first-identified NMDAR channel subunit, GluN1, was cloned and functionally expressed using a *Xenopus* oocyte expression system in 1991 (Moriyoshi et al. 1991). Subsequently, eight splice variants of the GluN1 subunit have been identified (Yamazaki et al. 1992; Sugihara et al. 1992; Hollmann et al. 1993). These splice variants show different spatiotemporal expression patterns, suggesting different regulatory mechanisms for splicing (Zhong et al. 1994; Laurie and Seeburg 1994). The GluN1 splice variants have four different cytoplasmic carboxy-terminal domains (CTD) derived from alternative splicing. These four CTDs are involved in functional modification, transport, surface expression, and membrane localization of the GluN1 subunit (Lau and Zukin 2007). The CTD variants are also involved in the interaction of NMDAR with many postsynaptic scaffold proteins and signaling molecules (Husi et al. 2000). The ligand-binding domain (LBD) of GluN1 recognizes D-serine or glycine, but not glutamate.

Members of a second NMDAR subunit subfamily (GluN2A, GluN2B, GluN2C, and GluN2D) were subsequently identified (Meguro et al. 1992; Kutsuwada et al. 1992; Ikeda et al. 1992; Monyer et al. 1992; Ishii et al. 1993). The GluN2 subunits recognize glutamate. Combining GluN1 with one member of the GluN2 subfamily can reconstitute highly active NMDAR channels, suggesting that active NMDAR is composed of GluN1 and GluN2 heteromers. Native NMDARs include at least two GluN1 subunits and two GluN2 subunits. Thus, NMDAR is an assembly of GluN1 and GluN2 subunits (GluN1/GluN2) that constitute a dimer of heterodimers (heterotetramer) (Furukawa et al. 2005; Traynelis et al. 2010; Salussolia et al. 2011). A triheteromeric tetramer composed of two GluN1 subunits, one GluN2A subunit, and one GluN2B subunit (GluN1/GluN2A/GluN2B) has also been observed (Sheng et al. 1994). The molecular cloning of four distinct GluN2 subunits reveals the molecular diversity of NMDARs. The properties of NMDARs composed of different combinations of GluN1 and GluN2, such as their affinities for ligands, sensitivities to antagonists and modulators, and modulations of channel properties, are dependent on the GluN2 subunit. The intracellular CTD of the GluN2 subunit is also involved in interactions of the subunit with many postsynaptic scaffold proteins and signaling molecules. The GluN2A subunit has two alternatively spliced carboxy termini (Salussolia et al. 2011).

A third subfamily of NMDAR subunits (composed of GluN3A and GluN3B) was identified by PCR-based homology cloning (Ciabarra et al. 1995; Sucher et al. 1995; Matsuda et al. 2002). GluN3 subunits recognize glycine or D-serine. The combination of GluN1, GluN3A, and GluN3B constitutes glycine-gated cation channels in mammalian cells (Smothers and Woodward 2007). In combination with GluN1 and GluN2 subunits, the GluN3 subunit decreases NMDAR channel activity. Thus, GluN3 is considered the inhibitory subunit of NMDAR (Smothers and Woodward 2007). The coexpression of GluN3B with GluN1/GluN2A and GluN1/GluN2B heteromeric channels in *Xenopus* oocytes does not alter their sensitivities to Mg^{2+}, ketamine, isoflurane, NO, or ethanol (Yamakura et al. 2005). Recently, functional GluN1/GluN3A heteromeric channels are identified in mouse medial habenula as an excitatory glycinergic NMDARs controlling of negatively valued emotional associations (Otsu et al. 2019).

3.2 Distribution of NMDAR Subunits

Knowledge of NMDAR localization in vivo is informative in the study of systemic ketamine treatment. The spatiotemporal expression patterns of each NMDAR subunit suggest the functional diversity of NMDARs in vivo. The GluN1 subunit is expressed ubiquitously in the CNS from the embryonic to adult stages (Watanabe et al. 1992, 1993). In the embryonic mouse CNS, GluN2B is expressed ubiquitously, and GluN2D is predominantly expressed in the brainstem (Watanabe et al. 1992, 1993). After birth, the expression of GluN2B is restricted to the forebrain, and expression levels of the GluN2D subunit markedly decrease. The expression of GluN2A is ubiquitous in the brain and enhanced after birth, and expression of GluN2C is restricted mainly to cerebellar granule cells. In the adult forebrain, triheteromeric NMDAR (GluN1/GluN2A/GluN2B) has been identified (Sheng et al. 1994). Microscale analyses revealed distinct trafficking mechanism and functions of the GluN1/GluN2A and GluN1/GluN2B NMDARs to synaptic and extrasynaptic regions (Hardingham and Bading 2010; Zhang et al. 2015). The expression levels of GluN3A are higher at younger stages and decrease later in life in many brain regions (Wong et al. 2002). Expression levels of GluN3B are higher at younger stages of life than later and are detected specifically in the pons, midbrain, medulla, and spinal cord (Matsuda et al. 2002). The molecular mechanisms regulating these unique expression patterns of NMDAR subunits have not been clarified. In addition to expression of NMDAR subunits in the CNS, some NMDAR subunits have been detected in peripheral tissues, such as the islets of Langerhans in the pancreas (Moriyama and Hayashi 2003; Marquard et al. 2015), heart (Gill et al. 1998), bone (Szczesniak et al. 2005), and cancer cells (Li and Hanahan 2013).

3.3 Structural and Functional Domains of NMDARs

GluR subunits possess a four-part modular domain structure. The amino-terminal domain (ATD) and the LBD reside in the extracellular portion of the receptor. The transmembrane domain (TMD) contains four hydrophobic membrane regions (M1 to M4) and defines the ion channel pore that is the site of action for open channel blockers. Finally, there is an intracellular CTD.

3.3.1 ATD

Members of the GluR channel family have 400- to 450-amino-acid-long ATDs (Fig. 2). The ATDs of NMDAR subunits are not essential for channel activity but are involved in the control of pharmacological and kinetic properties (Yuan et al. 2009). The ATD has a clamshell-like structure composed of two R1 and R2 domains (Karakas et al. 2009). The ATDs of GluN2A and GluN2B possess the sites for

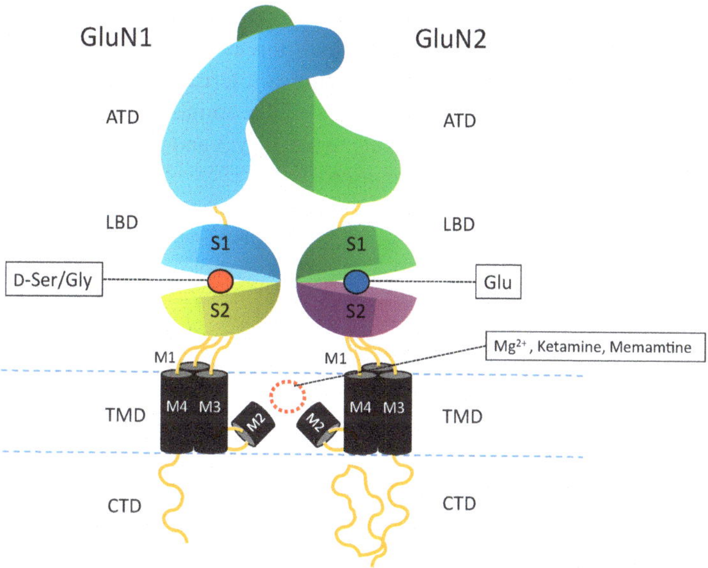

Fig. 2 Schematic structures of heteromeric NMDARs. Schematic structures of heterodimeric GluN1/GluN2 NMDAR channels. Extracellular amino-terminal domains (ATDs), ligand-binding domains (LBDs), transmembrane domains (TMDs), and intracellular carboxy-terminal domains (CTDs) are indicated. Locations of the binding sites of endogenous ligands (glutamate and D-serine/glycine), Mg^{2+}, and open channel blockers are shown. LBDs is formed by two extracellular structures referred as S1 and S2. TMDs contain four hydrophobic membrane regions M1 to M4

inhibitory Zn^{2+} binding and GluN2B-specific ifenprodil binding, respectively (Karakas et al. 2009).

3.3.2 LBD

The endogenous neurotransmitter agonist for NMDARs is glutamate, but the full activation of NMDAR requires concomitant binding of glutamate with a co-agonist, glycine or D-serine (Kleckner and Dingledine 1988), to the LBDs of NMDAR. Glutamate binds to the GluN2 subunit, and glycine or D-serine binds to the GluN1 subunit. Thus, both the GluN1 and GluN2 subunits are essential for activation of NMDARs. Although glycine is abundant in the CNS, the co-agonist-binding site of NMDARs is not saturated with glycine because extracellular glycine is actively taken up by highly expressed glycine transporters in the CNS (Zafra et al. 1995). A significant amount of free D-serine, which cannot be used for protein synthesis, has been detected in the forebrain (Hashimoto et al. 1992). D-Serine is produced by the enzyme serine racemase (SRR) in the CNS (Wolosker et al. 1999). Enzymatic degradation of D-serine reduces both spontaneous and evoked NMDAR currents (Mothet et al. 2000), suggesting that D-serine is the predominant endogenous co-agonist of NMDAR. The differential roles of the two co-agonists, glycine, and D-serine in the regulation of NMDARs have been discussed extensively (Mothet et al. 2015). The LBD is highly conserved in different GluR families, and the LBD of GluR is formed by two extracellular structures referred to as S1 and S2 (Stern-Bach et al. 1994). S1 is located on the extracellular amino-terminal side near M1, and S2 is located on the extracellular side between M3 and M4 (Fig. 2). Structures of LBD show a clamshell-like conformation.

3.3.3 TMD

NMDAR channels have unique properties compared with non-NMDAR channels. The activation of NMDAR requires the binding of two ligands (an agonist and co-agonist) and release from extracellular Mg^{2+} blocking in the channel pore. The positive charge of Mg^{2+} causes Mg^{2+} blocking of NMDAR by the electronic force of the negative charge of a neuron under the physiological membrane potential of about −70 mV. NMDAR is activated by the depolarization of membrane potential. Thus, NMDARs are ligand-gated and voltage-dependent ion channels. Both properties are necessary for detecting the coincidence of presynaptic excitation (glutamate release) and postsynaptic activation (membrane depolarization). In order for this coincidental detection to function properly, Mg^{2+} blocking is critical. With the opening of NMDAR ion channels, the monovalent cations Na^+ and K^+ can enter and permeate the channel, depending on their electronic and concentration gradients. In contrast to non-NMDAR channels, Ca^{2+} can selectively enter NMDAR channels and induce intracellular Ca^{2+}-dependent signal transduction in neurons. A low level of Ca^{2+} influx into neurons is necessary for the survival of neurons; however, excessive Ca^{2+} influx induces neuronal damage and death.

The process by which agonist binding leads to channel opening of NMDARs consists of three sequential steps: (1) agonist binding; (2) conformational changes such as the clam shell closure of the LBD; and (3) further conformational change of the ion channel pore to an open configuration. Introduction of mutations in the second hydrophobic region (M2) of the TMD affects ion channel properties. Thus, M2 is critical in the formation of the ion channel pore. M2 possesses a critical asparagine (N) residue that determines Mg^{2+} blocking and permeability to Ca^{2+} (Mori et al. 1992; Burnashev et al. 1992) (Fig. 3). The replacement of this N residue by glutamine (Q) or arginine (R) was shown to reduce sensitivity to Mg^{2+} and the open channel blocker MK-801, suggesting that the binding site of MK-801 overlaps with the Mg^{2+}-binding site (Mori et al. 1992). Subsequent analyses suggested that the topology of NMDAR subunits has a structure of three transmembrane helices (M1, M3, and M4) and one loop (M2) (Fig. 2), and the M2 loop lines the inner cavity of the ion channel pore. The functionally critical N residue in M2 of NMDAR is also involved in channel blocking by NMDAR-specific open channel blockers (ketamine, PCP, MK-801, and memantine, as shown in Fig. 1), polyamines, and protons (Yamakura et al. 1993; Kashiwagi et al. 2002).

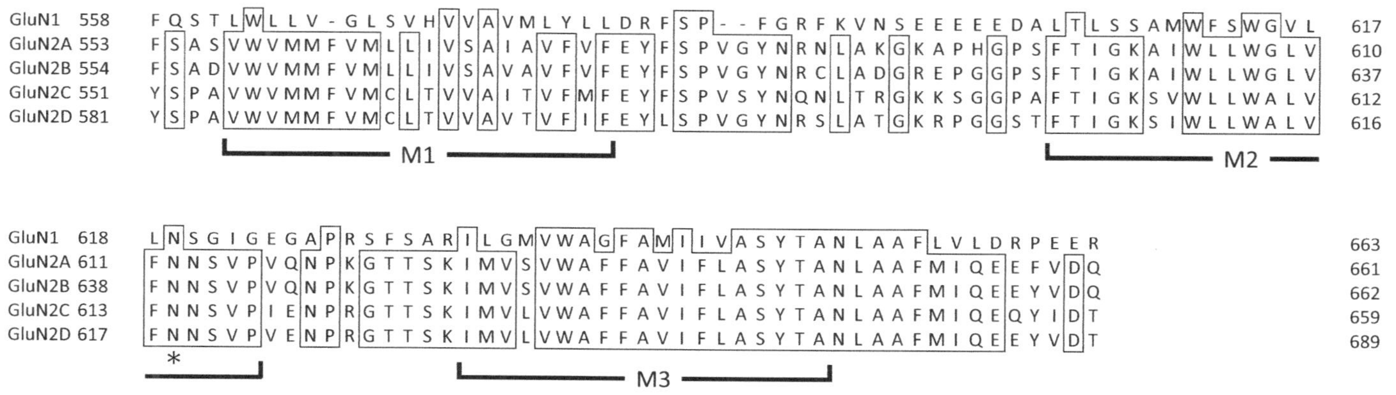

Fig. 3 Amino acid sequence alignment of M1 to M3 of the human GluN1, GluN2A, GluN2B, GluN2C, and GluN2D subunits. Amino acid sequences of human NMDAR subunits (GluN1, GluN2A, GluN2B, GluN2C, and GluN2D) are shown. M1 to M3 are identified by their Kyte and Doolittle hydropathy profiles. In M2, critical and conserved amino acid residue asparagine (N) determining Mg^{2+} blocking, Ca^{2+} permeability, and sensitivity to channel blockers is indicated with asterisk (∗). Sets of identical amino acid residues among five subunits or four GluN2 subunits are enclosed. Amino-acid sequence data reported are available in the GenBank databases under the accession numbers (GluN1, NP_015566.1; GluN2A, NP_000824.1; GluN2B, NP_000825.2; GluN2C, NP_000826.2; GluN2D, NP_000827.2)

3.3.4 CTD

The GluN1 subunit of NMDAR can have four distinct carboxy-termini derived from alternative splicing, and the GluN2 subunit has a long carboxy-terminal region. The CTD of NMDAR affects membrane targeting, stabilization, and degradation (Vissel et al. 2001; Hayashi et al. 2009) and is modified by phosphorylation and palmitoylation. The CTD of NMDAR also provides interaction sites for many intracellular proteins important for signal transduction and synaptic formation and is involved in the formation of NMDAR complexes at the postsynaptic site (Husi et al. 2000).

3.3.5 Crystal Structure of NMDAR with Open Channel Blockers

The expression of intact heteromeric NMDARs is challenging because of the technical difficulties associated with overexpression, purification, and assembly of heterooligomers. With many methodological improvements in X-ray crystallography and electron cryo-microscopy (cryo-EM), studies have revealed the structures of the NMDAR subunit domains of *Xenopus* and rat (Karakas and Furukawa 2014; Lee et al. 2014). The sites of action for ketamine, MK-801, and memantine might all overlap in the TMD, as described previously. Neither crystal nor cryo-EM structures of heteromeric NMDARs in complex with ketamine have been reported, likely due to the low affinity of ketamine for NMDARs. In contrast, cryo-EM density maps and electron density maps from X-ray crystallography of the TMD have identified the site and mode of MK-801 and memantine binding (Song et al. 2018). The M3 helix is a critical structural determinant linking agonist binding to channel opening. The M3, M2, and the pore loop (P-loop) have shown well-defined densities (Lu et al. 2017). A complex with MK-801, the M3 helix, and the tip of the P-loop forms a pyramidal vestibule harboring an elliptical density feature representing a trapped MK-801 molecule, with its two aromatic rings positioned toward the M3 helices of the GluN2 subunits (Song et al. 2018; Lu et al. 2017). The amine group of MK-801 formed stable hydrogen bonds with the two P-loop asparagine (N) residues in M2 (Song et al. 2018) (Fig. 4). Memantine also binds at a location nearly identical to the binding site of MK-801. Because of the 100-fold lower affinity of memantine for NMDAR compared with MK-801, fine mapping of the ketamine-binding site in the crystal structure of NMDAR is expected in the future.

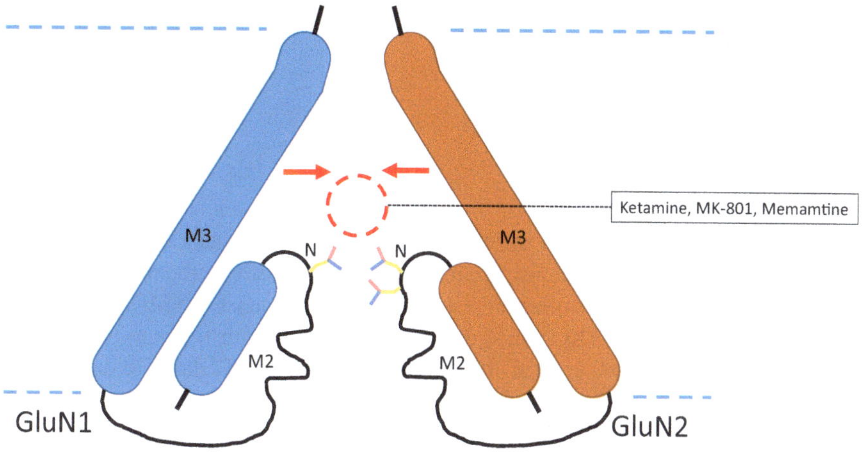

Fig. 4 Schematic representation of open channel blocking of NMDARs. Representation of the putative ketamine-binding site in the pore and vestibular regions based on the crystal structures of NMDAR with MK-801 and memantine modified from (Song et al. 2018). Loop (line) and helix (bar) in M2 and helix in M3 are indicated with the critical asparagine (N) residues. Channel blockers induce channel closure (arrows)

4 Site of Ketamine Action on NMDARs

4.1 *In Vitro Studies*

The effects of ketamine and other open channel blockers on the activity of recombinant heteromeric NMDAR channels expressed in *Xenopus* oocytes and mammalian cells have been examined. The sensitivities to ketamine and PCP of four NMDAR channels (GluN1/GluN2A, GluN1/GluN2B, GluN1/GluN2C, and GluN1/GluN2D) expressed in *Xenopus* oocytes are similar (Yamakura et al. 1993). The GluN1/GluN2A and GluN1/GluN2B channels were more sensitive to MK-801 than the GluN1/GluN2C and GluN1/GluN2D channels. Because the affinity for MK-801 is higher than that for ketamine or PCP, the suppressive effects of ketamine and PCP on recombinant heteromeric NMDARs are reversible, but the effects of MK-801 are not. As mentioned above, the effects of mutation of a critical N residue in M2 on Mg^{2+} blocking, inhibition by ketamine, PCP, and MK-801 are similar. These results suggest that Mg^{2+}, PCP, ketamine, and MK-801 act at a common site on NMDARs. Furthermore, the suppressive effects of ketamine on recombinant heteromeric NMDARs expressed in mammalian cells are dependent on the presence of Mg^{2+}. At an extracellular Mg^{2+} concentration of 1 mM, ketamine concentration–inhibition curves of recombinant heteromeric NMDARs were right shifted. Under these conditions, the GluN1/GluN2C channel is the most sensitive to ketamine among GluN1/GluN2 channels (Kotermanski and Johnson 2009). The sensitivity to memantine, another NMDAR open channel blocker, is also affected by the presence of Mg^{2+}. These results suggest that (1) the sensitivity of NMDAR blocking sites to Mg^{2+}, ketamine, PCP, and memantine is dependent on the constituent GluN2 subunit and (2) these blocking sites are overlapped.

4.2 *In Vivo Studies*

Antidepressant phenotypes and the effects of ketamine have been examined using NMDAR gene-manipulated mouse strains. Because the GluN1 subunit is expressed in the entire brain from the embryonic stage and is necessary for NMDAR channel activity, GluN1-knockout (KO) mice die immediately after birth due to respiratory failure (Forrest et al. 1994; Li et al. 1994). The contribution of GluN1 to depressive or antidepressive phenotypes is unknown. Cre-recombinase-mediated conditional (spatiotemporally selective) GluN1-KO mice have been established (Tsien et al. 1996; McHugh et al. 1996; Iwasato et al. 2000; Nakazawa et al. 2002), and conditional KO strains will be useful for assessing behavioral effects of ketamine treatment.

GluN2A-KO mice are viable and grow normally but show impaired synaptic plasticity, spatial learning, and motor learning in the adult stage (Sakimura et al. 1995; Kiyama et al. 1998; Kishimoto et al. 2001). Furthermore, GluN2A-KO mice show some schizophrenia-like phenotypes (Miyamoto et al. 2001), supporting the hypo-NMDAR function hypothesis of schizophrenia. The effect of ketamine treatment on GluN2A-KO mice was evaluated with a righting reflex test. The GluN2A-mutant mice were more resistant to ketamine than control mice, suggesting involvement of the GluN2A subunit in ketamine sensitivity in vivo (Petrenko et al. 2004). Furthermore, GluN2A-KO mice are resistant to inflammation-induced depression (Francija et al. 2019).

The GluN2B subunit is predominantly expressed throughout the brain at embryonic and neonatal stages. GluN2B-KO mice die immediately after birth because of impaired neural network formation and loss of the suckling response (Kutsuwada et al. 1996). Principal cortical neuron-specific GluN2B conditional KO mice showed decreased despair-like (antidepressive) behaviors and showed no effects from ketamine treatment (Miller et al. 2014). Furthermore, treatment of wild-type mice with the GluN2B-selective antagonist Ro 25-6981 induced an antidepressive state (Miller et al. 2014). These findings support the "direct inhibition hypothesis" of the antidepressive effects of ketamine, as described later.

As described above, GluN2C-containing heteromeric recombinant NMDAR channels in vitro show higher affinity to ketamine under the more physiological condition of 1 mM Mg^{2+} (Kotermanski and Johnson 2009). GluN2C-KO mice show hypo-NMDAR and schizophrenia-related behaviors, but the effects of ketamine treatment have not been reported (Hillman et al. 2011).

Expression of GluN2D has been observed in GABAergic interneurons, suggesting a distinct function of this subunit (Yamasaki et al. 2014). GluN2D-KO mice show normal development and breeding but show some indications of an impaired emotional state (Ikeda et al. 1995; Miyamoto et al. 2002). PCP-induced hyperlocomotion, motor impairment, and increased dopamine levels were not observed in GluN2D-KO mice (Yamamoto et al. 2013; Hagino et al. 2010). Ketamine-induced locomotor sensitization, increased brain activation (detected with the uptake of 2-deoxyglucose), and increased cortical gamma-band oscillatory power were also not observed in GluN2D-KO mice (Sapkota et al. 2016; Yamamoto et al. 2016). Ketamine is an arylcycloalkylamine that exists as a racemic mixture of S and R isomers (Fig. 1). Purified (S)-ketamine has a higher affinity to NMDAR than the R isomer (Hashimoto 2019). Interestingly, (R)-ketamine induces sustained antidepressive effects in control mice but not in GluN2D-KO mice (Ide et al. 2017, 2019). These data suggest a critical role for the GluN2D subunit and its expressing neurons in the antidepressive effects of ketamine.

As mentioned above, D-serine produced by SRR is an endogenous co-agonist of NMDARs. SRR-KO mice showed reduced D-serine (about 10% of control) in forebrain regions and hypo-NMDAR functional states (Inoue et al. 2008; Basu et al. 2009). Although the effects of ketamine in SRR-KO mice were not examined, SRR-KO mice showed antidepressive phenotypes against chronic social defeat stress (Dong et al. 2018).

The above analyses using NMDAR-specific and -related gene-manipulated mouse strains suggest (1) differential roles of GluN2 subunits and their expressing neurons in ketamine action; (2) hypo-NMDAR states induce schizophrenia-like behavioral changes like ketamine treatment; and (3) some hypo-NMDAR states are related with antidepressive phenotypes, as with ketamine-induced antidepressive phenotypes.

4.3 *Possible Network Mechanisms of Antidepressant Effect of Ketamine*

Single- and low-dose ketamine treatments result in fast-onset and continuous antidepressant effect. Ketamine treatment induces activation in the prefrontal cortex (PFC) of the brain. fMRI and FDG-PET studies have revealed accumulation of FDG, suggesting activation of the brain regions with the treatment. In experimental animals, antidepressive phenotypes induced by ketamine treatment are associated with an increase in translation of proteins, including brain-derived neurotrophic factor (Autry et al. 2011).

There are two hypotheses for the mechanisms of action of ketamine at the neural level (Miller et al. 2016). The first is called the "direct hypothesis" because the direct inhibition of NMDARs on cortical pyramidal (excitatory) neurons by ketamine is thought to induce homeostatic and compensatory activation of the excitation. This hypothesis is supported by evidence from a model mouse in which the GluN2B subunit is selectively removed in cortical pyramidal neurons (Miller et al. 2014). One target brain region for the antidepressant effects of ketamine is the PFC. This mouse model showed enhanced protein synthesis rates and increased excitatory drive in pyramidal neurons of the PFC. Furthermore, this mouse model mimics and occludes the behavioral action of ketamine. Infusion of a GluN2B-selective NMDAR antagonist into the PFC also mimics the antidepressant effects of ketamine. In contrast, the "indirect hypothesis" proposes that low doses of ketamine selectively suppress NMDARs on cortical inhibitory interneurons, leading to disinhibition and indirect excitation of excitatory pyramidal neurons. This then leads to protein synthesis, activity-dependent synaptic plasticity, and increased excitatory synaptic drive. This hypothesis is supported with the evidence that parvalbumin-positive inhibitory interneurons in the PFC are fast-spiking (high frequency of action potentials) and NMDARs on interneurons are tonically active and more sensitive to the NMDAR open channel blocker ketamine than excitatory pyramidal neurons (Li et al. 2002). In addition to the findings mentioned above, there are many studies both supporting and contradicting each of these hypotheses (Miller et al. 2016). Furthermore, a critical role for neural networks in the lateral habenular has been proposed in the antidepressive effects of ketamine (Cui et al. 2019).

5 Other Possibilities

5.1 Target Molecules of Ketamine Other Than NMDARs

In addition to being an NMDAR open channel blocker, ketamine also acts on other targets such as L-type voltage-dependent Ca^{2+} channels (Yamakage et al. 1995), hyperpolarization-activated cyclic nucleotide channels (Chen et al. 2009; Zhou et al. 2013), the serotonin reuptake system (Martin et al. 1982), and large-conductance K_{Ca} channels (BK channels) (Hayashi et al. 2011). The possible roles of these targets in the antidepressive effects of ketamine have not been resolved at present.

5.2 Effects of Ketamine Metabolites

(R,S)-ketamine is metabolized to (2S,6S;2R,6R)-hydroxynorketamine (HNK) in vivo. One of the enantiomers, (2R,6R)-HNK, showed antidepressant effects in mice. These antidepressant effects are independent of NMDAR blocking but involve early and sustained activation of AMPA receptors (Zanos et al. 2016). However, the antidepressive effects of (2R,6R)-HNK were not observed in chronic social defeat stress and lipopolysaccharide-induced depression models (Yang et al. 2017).

6 Conclusions and Perspectives

Ketamine treatment results in a fast-onset and continuous antidepressant effect. The main targets of ketamine are NMDARs. NMDARs in the brain are spatiotemporal heterogenous populations. Furthermore, the functions of NMDARs are modulated by many endogenous and exogenous substances. Fine crystal or cryo-EM structure of ketamine-NMDAR complex is yet to be determined. Information from fine crystal and dynamic ketamine-NMDAR structures will potentially lead to the development of novel antidepressants for therapeutic applications.

Acknowledgments I thank Dr. Hironori Izumi for the preparation of the figures. Parts of this work were supported by a Grant from the Ministry of Education, Culture, Sports, Science and Technology (MEXT) of Japan (KAKENHI, Grant No. 18K06888).

Conflict of interest: The author declares no conflicts of interest with the content of this chapter.

References

Anis NA, Berry SC, Burton NR, Lodge D (1983) The dissociative anaesthetics, ketamine and phencyclidine, selectively reduce excitation of central mammalian neurones by N-methyl-aspartate. Br J Pharmacol 79:565–575

Autry AE, Adachi M, Nosyreva E, Na ES, Los MF, Cheng PF, Kavalali ET, Monteggia LM (2011) NMDA receptor blockade at rest triggers rapid behavioural antidepressant responses. Nature 475:91–95

Basu AC, Tsai GE, Ma CL, Ehmsen JT, Musutafa AK, Han L, Jiang ZI, Benneyworth MA, Froimowitz MP, Lange N, Snyder SH, Bergeron R, Coyle JT (2009) Targeted disruption of serine racemase affects glutamatergic neurotransmission and behavior. Mol Psychiatry 14:719–727

Bergman SA (1999) Ketamine: review of its pharmacology and its use in pediatric anesthesia. Anesth Prog 46:10–20

Berman RM, Cappiello A, Anand A, Oren DA, Heninger GR, Charney DS, Krystal JH (2000) Antidepressant effects of ketamine in depressed patients. Biol Psychiatry 47:351–354

Burnashev N, Schoepfer R, Monyer H, Ruppersberg JP, Günther W, Seeburg PH, Sakmann B (1992) Control by asparagine residues of calcium permeability and magnesium blockade in the NMDA receptor. Science 257:1415–1419

Chen X, Shu S, Bayliss DA (2009) HCN1 channel subunits are a molecular substrate for hypnotic actions of ketamine. J Neurosci 29:600–609

Ciabarra AM, Sullivan JM, Gahn LG, Pecht G, Heinemann S, Sevarino KA (1995) Cloning and characterization of chi-1: a developmentally regulated member of a novel class of the ionotropic glutamate receptor family. J Neurosci 15:6498–6508

Collingridge GL, Kehl SJ, McLennan H (1983) Excitatory amino acids in synaptic transmission in the Schaffer collateral-commissural pathway of the rat hippocampus. J Physiol 334:33–46

Cui Y, Hu S, Hu H (2019) Lateral habenular burst firing as a target of the rapid antidepressant effects of ketamine. Trends Neurosci 42:179–191

Curtis DR, Watkins JC (1963) Acidic amino acids with strong excitatory actions on mammalian neurones. J Physiol 166:1–14

De Simoni S, Schwarz AJ, O'Daly OG, Marquand AF, Brittain C, Gonzales G, Stephenson S, Williams SC, Mehta MA (2013) Test-retest reliability of the BOLD pharmacological MRI response to ketamine in healthy volunteers. Neuroimage 64:75–90

Domino EF, Domino SE, Smith RE, Domino LE, Goulet JR, Domino KE, Zsigmond EK (1984) Ketamine kinetics in unmedicated and diazepam-premedicated subjects. Clin Pharmacol Ther 36:645–653

Dong C, Zhang JC, Ren Q, Ma M, Qu Y, Zhang K, Yao W, Ishima T, Mori H, Hashimoto K (2018) Deletion of serine racemase confers D-serine -dependent resilience to chronic social defeat stress. Neurochem Int 116:43–51

Forrest D, Yuzaki M, Soares HD, Ng L, Luk DC, Sheng M, Stewart CL, Morgan JI, Connor JA, Curran T (1994) Targeted disruption of NMDA receptor 1 gene abolishes NMDA response and results in neonatal death. Neuron 13:325–338

Francija E, Petrovic Z, Brkic Z, Mitic M, Radulovic J, Adzic M (2019) Disruption of the NMDA receptor GluN2A subunit abolishes inflammation-induced depression. Behav Brain Res 359:550–559

Furukawa H, Singh SK, Mancusso R, Gouaux E (2005) Subunit arrangement and function in NMDA receptors. Nature 438:185–192

Garfield JM, Garfield FB, Stone JG, Hopkins D, Johns LA (1972) A comparison of psychologic responses to ketamine and thiopental-nitrous oxide-halothane anesthesia. Anesthesiology 36:329–338

Gill SS, Pulido OM, Mueller RW, McGuire PF (1998) Molecular and immunochemical characterization of the ionotropic glutamate receptors in the rat heart. Brain Res Bull 46:429–434

Graifenstein FF, Devault M, Yoshitake J, Gejewski JE (1958) A study of a 1-aryl cyclo hexyl amine for anesthesia. Anesth Analg 37:283–294

Hagino Y, Kasai S, Han W, Yamamoto H, Nabeshima T, Mishina M, Ikeda K (2010) Essential role of NMDA receptor channel ε4 subunit (GluN2D) in the effects of phencyclidine, but not methamphetamine. PLoS One 5:e13722

Hardingham GE, Bading H (2010) Synaptic versus extrasynaptic NMDA receptor signalling: implications for neurodegenerative disorders. Nat Rev Neurosci 11:682–696

Hashimoto K (2019) Rapid-acting antidepressant ketamine, its metabolites and other candidates: a historical overview and future perspective. Psychiatry Clin Neurosci. https://doi.org/10.1111/pcn.12902

Hashimoto A, Nishikawa T, Hayashi T, Fujii N, Harada K, Oka T, Takahashi K (1992) The presence of free D-serine in rat brain. FEBS Lett 296:33–36

Hayashi T, Thomas GM, Huganir RL (2009) Dual palmitoylation of NR2 subunits regulates NMDA receptor trafficking. Neuron 64:213–226. https://doi.org/10.1016/j.neuron.2009.08.017

Hayashi Y, Kawaji K, Sun L, Zhang X, Koyano K, Yokoyama T, Kohsaka S, Inoue K, Nakanishi H (2011) Microglial Ca^{2+}-activated K^+ channels are possible molecular targets for the analgesic effects of S-ketamine on neuropathic pain. J Neurosci 31:17370–17382

Hillman BG, Gupta SC, Stairs DJ, Buonanno A, Dravid SM (2011) Behavioral analysis of NR2C knockout mouse reveals deficit in acquisition of conditioned fear and working memory. Neurobiol Learn Mem 95:404–414

Hollmann M, Boulter J, Maron C, Beasley L, Sullivan J, Pecht G, Heinemann S (1993) Zinc potentiates agonist-induced currents at certain splice variants of the NMDA receptor. Neuron 10:943–954

Husi H, Ward MA, Choudhary JS, Blackstock WP, Grant SG (2000) Proteomic analysis of NMDA receptor-adhesion protein signaling complexes. Nat Neurosci 3:661–669

Ide S, Ikekubo Y, Mishina M, Hashimoto K, Ikeda K (2017) Role of NMDA receptor GluN2D subunit in the antidepressant effects of enantiomers of ketamine. J Pharmacol Sci 135:138–140

Ide S, Ikekubo Y, Mishina M, Hashimoto K, Ikeda K (2019) Cognitive impairment that is induced by (R)-ketamine is abolished in NMDA GluN2D receptor subunit knockout mice. Int J Neuropsychopharmacol 22:449–452

Ikeda K, Nagasawa M, Mori H, Araki K, Sakimura K, Watanabe M, Inoue Y, Mishina M (1992) Cloning and expression of the epsilon 4 subunit of the NMDA receptor channel. FEBS Lett 313:34–38

Ikeda K, Araki K, Takayama C, Inoue Y, Yagi T, Aizawa S, Mishina M (1995) Reduced spontaneous activity of mice defective in the epsilon 4 subunit of the NMDA receptor channel. Brain Res Mol Brain Res 33:61–71

Inoue R, Hashimoto K, Harai T, Mori H (2008) NMDA- and beta-amyloid1-42-induced neurotoxicity is attenuated in serine racemase knock-out mice. J Neurosci 28:14486–14491

Ishii T, Moriyoshi K, Sugihara H, Sakurada K, Kadotani H, Yokoi M, Akazawa C, Shigemoto R, Mizuno N, Masu M, Nakanishi S (1993) Molecular characterization of the family of the *N*-methyl-D-aspartate receptor subunits. J Biol Chem 268:2836–2843

Iwasato T, Datwani A, Wolf AM, Nishiyama H, Taguchi Y, Tonegawa S, Knöpfel T, Erzurumlu RS, Itohara S (2000) Cortex-restricted disruption of NMDAR1 impairs neuronal patterns in the barrel cortex. Nature 406:726–731

Johnstone M, Evans V, Baigel S (1959) Sernyl (CI-395) in clinical anaesthesia. Br J Anaesth 31:433–439

Karakas E, Furukawa H (2014) Crystal structure of a heterotetrameric NMDA receptor ion channel. Science 344:992–997. https://doi.org/10.1126/science.1251915

Karakas E, Simorowski N, Furukawa H (2009) Structure of the zinc-bound amino-terminal domain of the NMDA receptor NR2B subunit. EMBO J 28:3910–3920. https://doi.org/10.1038/emboj.2009.338

Kashiwagi K, Masuko T, Nguyen CD, Kuno T, Tanaka I, Igarashi K, Williams K (2002) Channel blockers acting at *N*-methyl-D-aspartate receptors: differential effects of mutations in the vestibule and ion channel pore. Mol Pharmacol 61:533–545

Kehrer C, Maziashvili N, Duglandze T, Gloveli T (2008) Altered excitatory-inhibitory balance in the NMDA-hypofunction model of schizophrenia. Front Mol Neurosci 1:6

Kishimoto Y, Kawahara S, Mori H, Mishina M, Kirino Y (2001) Long-trace interval eyeblink conditioning is impaired in mutant mice lacking the NMDA receptor subunit epsilon 1. Eur J Neurosci 13:1221–1227

Kiyama Y, Manabe T, Sakimura K, Kawakami F, Mori H, Mishina M (1998) Increased thresholds for long-term potentiation and contextual learning in mice lacking the NMDA-type glutamate receptor epsilon1 subunit. J Neurosci 18:6704–6712

Kleckner NW, Dingledine R (1988) Requirement for glycine in activation of NMDA-receptors expressed in Xenopus oocytes. Science 241:835–837

Kohrs R, Durieux ME (1998) Ketamine: teaching an old drug new tricks. Anesth Analg 87:1186–1193

Kotermanski SE, Johnson JW (2009) Mg^{2+} imparts NMDA receptor subtype selectivity to the Alzheimer's drug memantine. J Neurosci 29:2774–2779

Kutsuwada T, Kashiwabuchi N, Mori H, Sakimura K, Kushiya E, Araki K, Meguro H, Masaki H, Kumanishi T, Arakawa M, Mishina M (1992) Molecular diversity of the NMDA receptor channel. Nature 358:36–41

Kutsuwada T, Sakimura K, Manabe T, Takayama C, Katakura N, Kushiya E, Natsume R, Watanabe M, Inoue Y, Yagi T, Aizawa S, Arakawa M, Takahashi T, Nakamura Y, Mori H, Mishina M (1996) Impairment of suckling response, trigeminal neuronal pattern formation, and hippocampal LTD in NMDA receptor epsilon 2 subunit mutant mice. Neuron 16:333–344

Lau CG, Zukin RS (2007) NMDA receptor trafficking in synaptic plasticity and neuropsychiatric disorders. Nat Rev Neurosci 8:413–426

Laurie DJ, Seeburg PH (1994) Regional and developmental heterogeneity in splicing of the rat brain NMDAR1 mRNA. J Neurosci 14:3180–3194

Lee CH, Lü W, Michel JC, Goehring A, Du J, Song X, Gouaux E (2014) NMDA receptor structures reveal subunit arrangement and pore architecture. Nature 511:191–197. https://doi.org/10.1038/nature13548

Li L, Hanahan D (2013) Hijacking the neuronal NMDAR signaling circuit to promote tumor growth and invasion. Cell 153:86–100. https://doi.org/10.1016/j.cell.2013.02.051

Li Y, Erzurumlu RS, Chen C, Jhaveri S, Tonegawa S (1994) Whisker-related neuronal patterns fail to develop in the trigeminal brainstem nuclei of NMDAR1 knockout mice. Cell 76:427–437

Li Q, Clark S, Lewis DV, Wilson WA (2002) NMDA receptor antagonists disinhibit rat posterior cingulate and retrosplenial cortices: a potential mechanism of neurotoxicity. J Neurosci 22:3070–3080

Lu W, Du J, Goehring A, Gouaux E (2017) Cryo-EM structures of the triheteromeric NMDA receptor and its allosteric modulation. Science 355:1282. https://doi.org/10.1126/science.aal3729

Marquard J, Otter S, Welters A, Stirban A, Fischer A, Eglinger J, Herebian D, Kletke O, Klemen MS, Stožer A, Wnendt S, Piemonti L, Köhler M, Ferrer J, Thorens B, Schliess F, Rupnik MS, Heise T, Berggren PO, Klöcker N, Meissner T, Mayatepek E, Eberhard D, Kragl M, Lammert E (2015) Characterization of pancreatic NMDA receptors as possible drug targets for diabetes treatment. Nat Med 21:363–372. https://doi.org/10.1038/nm.3822

Martin LL, Bouchal RL, Smith DJ (1982) Ketamine inhibits serotonin uptake in vivo. Neuropharmacology 21:113–118

Matsuda K, Kamiya Y, Matsuda S, Yuzaki M (2002) Cloning and characterization of a novel NMDA receptor subunit NR3B: a dominant subunit that reduces calcium permeability. Brain Res Mol Brain Res 100:43–52

McHugh TJ, Blum KI, Tsien JZ, Tonegawa S, Wilson MA (1996) Impaired hippocampal representation of space in CA1-specific NMDAR1 knockout mice. Cell 87:1339–1349

Meguro H, Mori H, Araki K, Kushiya E, Kutsuwada T, Yamazaki M, Kumanishi T, Arakawa M, Sakimura K, Mishina M (1992) Functional characterization of a heteromeric NMDA receptor channel expressed from cloned cDNAs. Nature 357:70–74

Miller OH, Yang L, Wang CC, Hargroder EA, Zhang Y, Delpire E, Hall BJ (2014) GluN2B-containing NMDA receptors regulate depression-like behavior and are critical for the rapid antidepressant actions of ketamine. eLife 3:e03581

Miller OH, Moran JT, Hall BJ (2016) Two cellular hypotheses explaining the initiation of ketamine's antidepressant actions: direct inhibition and disinhibition. Neuropharmacology 100:17–26

Miyamoto Y, Yamada K, Noda Y, Mori H, Mishina M, Nabeshima T (2001) Hyperfunction of dopaminergic and serotonergic neuronal systems in mice lacking the NMDA receptor epsilon1 subunit. J Neurosci 21:750–757

Miyamoto Y, Yamada K, Noda Y, Mori H, Mishina M, Nabeshima T (2002) Lower sensitivity to stress and altered monoaminergic neuronal function in mice lacking the NMDA receptor epsilon 4 subunit. J Neurosci 22:2335–2342

Monyer H, Sprengel R, Schoepfer R, Herb A, Higuchi M, Lomeli H, Burnashev N, Sakmann B, Seeburg PH (1992) Heteromeric NMDA receptors: molecular and functional distinction of subtypes. Science 256:1217–1221

Morgan CJ, Muetzelfeldt L, Curran HV (2009) Ketamine use, cognition and psychological wellbeing: a comparison of frequent, infrequent and ex-users with polydrug and non-using controls. Addiction 104:77–87

Mori H, Masaki H, Yamakura T, Mishina M (1992) Identification by mutagenesis of a Mg^{2+}-block site of the NMDA receptor channel. Nature 358:673–675

Moriyama Y, Hayashi M (2003) Glutamate-mediated signaling in the islets of Langerhans: a thread entangled. Trends Pharmacol Sci 24:511–517

Moriyoshi K, Masu M, Ishii T, Shigemoto R, Mizuno N, Nakanishi S (1991) Molecular cloning and characterization of the rat NMDA receptor. Nature 354:31–37

Morris RG, Anderson E, Lynch GS, Baudry M (1986) Selective impairment of learning and blockade of long-term potentiation by an *N*-methyl-D-aspartate receptor antagonist, AP5. Nature 319:774–776

Mothet JP, Parent AT, Wolosker H, Brady RO Jr, Linden DJ, Ferris CD, Rogawski MA, Snyder SH (2000) D-serine is an endogenous ligand for the glycine site of the *N*-methyl-D-aspartate receptor. Proc Natl Acad Sci U S A 97:4926–4931

Mothet JP, Le Bail M, Billard JM (2015) Time and space profiling of NMDA receptor co-agonist functions. J Neurochem 135:210–225. https://doi.org/10.1111/jnc.13204

Nakazawa K, Quirk MC, Chitwood RA, Watanabe M, Yeckel MF, Sun LD, Kato A, Carr CA, Johnston D, Wilson MA, Tonegawa S (2002) Requirement for hippocampal CA3 NMDA receptors in associative memory recall. Science 297:211–218

Olney JW, Newcomer JW, Farber NB (1999) NMDA receptor hypofunction model of schizophrenia. J Psychiatr Res 33:523–533

Otsu Y, Darcq E, Pietrajtis K, Mátyás F, Schwartz E, Bessaih T, Abi Gerges S, Rousseau CV, Grand T, Dieudonné S, Paoletti P, Acsády L, Agulhon C, Kieffer BL, Diana MA (2019) Control of aversion by glycine-gated GluN1/GluN3A NMDA receptors in the adult medial habenula. Science 366:250–254

Petrenko AB, Yamakura T, Fujiwara N, Askalany AR, Baba H, Sakimura K (2004) Reduced sensitivity to ketamine and pentobarbital in mice lacking the N-methyl-D-aspartate receptor GluRepsilon1 subunit. Anesth Analg 99:1136–1149

Rush AJ, Trivedi MH, Wisniewski SR, Nierenberg AA, Stewart JW, Warden D, Niederehe G, Thase ME, Lavori PW, Lebowitz BD, McGrath PJ, Rosenbaum JF, Sackeim HA, Kupfer DJ, Luther J, Fava M (2006) Acute and longer-term outcomes in depressed outpatients requiring one or several treatment steps: a STAR*D report. Am J Psychiatry 163(11):1905–1917

Sakimura K, Kutsuwada T, Ito I, Manabe T, Takayama C, Kushiya E, Yagi T, Aizawa S, Inoue Y, Sugiyama H, Mishina M (1995) Reduced hippocampal LTP and spatial learning in mice lacking NMDA receptor epsilon 1 subunit. Nature 373:151–155

Salussolia CL, Prodromou ML, Borker P, Wollmuth LP (2011) Arrangement of subunits in functional NMDA receptors. J Neurosci 31:11295–11304. https://doi.org/10.1523/JNEUROSCI.5612-10.2011

Sapkota K, Mao Z, Synowicki P, Lieber D, Liu M, Ikezu T, Gautam V, Monaghan DT (2016) GluN2D N-Methyl-d-Aspartate receptor subunit contribution to the stimulation of brain activity and gamma oscillations by ketamine: implications for schizophrenia. J Pharmacol Exp Ther 356:702–711

Sheng M, Cummings J, Roldan LA, Jan YN, Jan LY (1994) Changing subunit composition of heteromeric NMDA receptors during development of rat cortex. Nature 368:144–147

Smothers CT, Woodward JJ (2007) Pharmacological characterization of glycine-activated currents in HEK 293 cells expressing *N*-methyl-D-aspartate NR1 and NR3 subunits. J Pharmacol Exp Ther 322:739–748

Song X, Jensen MQ, Jogini V, Stein RA, Lee CH, Mchaourab HS, Shaw DE, Gouaux E (2018) Mechanism of NMDA receptor channel block by MK-801 and memantine. Nature 556:515–519

Stern-Bach Y, Bettler B, Hartley M, Sheppard PO, O'Hara PJ, Heinemann SF (1994) Agonist selectivity of glutamate receptors is specified by two domains structurally related to bacterial amino acid-binding proteins. Neuron 13:1345–1357

Sucher NJ, Akbarian S, Chi CL, Leclerc CL, Awobuluyi M, Deitcher DL, Wu MK, Yuan JP, Jones EG, Lipton SA (1995) Developmental and regional expression pattern of a novel NMDA receptor-like subunit (NMDAR-L) in the rodent brain. J Neurosci 15:6509–6520

Sugihara H, Moriyoshi K, Ishii T, Masu M, Nakanishi S (1992) Structures and properties of seven isoforms of the NMDA receptor generated by alternative splicing. Biochem Biophys Res Commun 185:826–832

Szczesniak AM, Gilbert RW, Mukhida M, Anderson GI (2005) Mechanical loading modulates glutamate receptor subunit expression in bone. Bone 37:63–73

Traynelis SF, Wollmuth LP, McBain CJ, Menniti FS, Vance KM, Ogden KK, Hansen KB, Yuan H, Myers SJ, Dingledine R (2010) Glutamate receptor ion channels: structure, regulation, and function. Pharmacol Rev 62:405–496. https://doi.org/10.1124/pr.109.002451

Tsien JZ, Huerta PT, Tonegawa S (1996) The essential role of hippocampal CA1 NMDA receptor-dependent synaptic plasticity in spatial memory. Cell 87:1327–1338

Vissel B, Krupp JJ, Heinemann SF, Westbrook GL (2001) A use-dependent tyrosine dephosphorylation of NMDA receptors is independent of ion flux. Nat Neurosci 4:587–596

Watanabe M, Inoue Y, Sakimura K, Mishina M (1992) Developmental changes in distribution of NMDA receptor channel subunit mRNAs. Neuroreport 3:1138–1140

Watanabe M, Inoue Y, Sakimura K, Mishina M (1993) Distinct distributions of five *N*-methyl-D-aspartate receptor channel subunit mRNAs in the forebrain. J Comp Neurol 338:377–390

Wolosker H, Blackshaw S, Snyder SH (1999) Serine racemase: a glial enzyme synthesizing D-serine to regulate glutamate-*N*-methyl-D-aspartate neurotransmission. Proc Natl Acad Sci U S A 96:13409–13414

Wong HK, Liu XB, Matos MF, Chan SF, Pérez-Otaño I, Boysen M, Cui J, Nakanishi N, Trimmer JS, Jones EG, Lipton SA, Sucher NJ (2002) Temporal and regional expression of NMDA receptor subunit NR3A in the mammalian brain. J Comp Neurol 450:303–317

Yamakage M, Hirshman CA, Croxton TL (1995) Inhibitory effects of thiopental, ketamine and Propofol on voltage-dependent Ca^{2+} channels in porcine tracheal smooth muscle cells. Anesthesiology 83:1274–1282

Yamakura T, Mori H, Masaki H, Shimoji K, Mishina M (1993) Different sensitivities of NMDA receptor channel subtypes to non-competitive antagonists. Neuroreport 4:687–690

Yamakura T, Askalany AR, Petrenko AB, Kohno T, Baba H, Sakimura K (2005) The NR3B subunit does not alter the anesthetic sensitivities of recombinant N-methyl-D-aspartate receptors. Anaesth Analg 100:1687–1692

Yamamoto H, Kamegaya E, Sawada W, Hasegawa R, Yamamoto T, Hagino Y, Takamatsu Y, Imai K, Koga H, Mishina M, Ikeda K (2013) Involvement of the N-methyl-D-aspartate receptor GluN2D subunit in phencyclidine-induced motor impairment, gene expression, and increased Fos immunoreactivity. Mol Brain 6:56

Yamamoto T, Nakayama T, Yamaguchi J, Matsuzawa M, Mishina M, Ikeda K, Yamamoto H (2016) Role of the NMDA receptor GluN2D subunit in the expression of ketamine-induced behavioral sensitization and region-specific activation of neuronal nitric oxide synthase. Neurosci Lett 610:48–53

Yamasaki M, Okada R, Takasaki C, Toki S, Fukaya M, Matsume R, Sakimura K, Mishina M, Shirakawa T, Watanabe M (2014) Opposing role of NMDA receptor GluN2B and GluN2D in somatosensory development and maturation. J Neurosci 34:11534–11548

Yamazaki M, Mori H, Araki K, Mori KJ, Mishina M (1992) Cloning, expression and modulation of a mouse NMDA receptor subunit. FEBS Lett 300:39–45

Yang C, Qu Y, Abe M, Nozawa D, Chaki S, Hashimoto K (2017) (*R*)-ketamine shows greater potency and longer lasting antidepressant effects than its metabolite (*2R,6R*)-hydroxynorketamine. Biol Psychiatry 82:e43–e44

Yuan H, Hansen KB, Vance KM, Ogden KK, Traynelis SF (2009) Control of NMDA receptor function by the NR2 subunit amino-terminal domain. J Neurosci 29:12045–12058. https://doi.org/10.1523/JNEUROSCI.1365-09.2009

Zafra F, Aragón C, Olivares L, Danbolt NC, Giménez C, Storm-Mathisen J (1995) Glycine transporters are differentially expressed among CNS cells. J Neurosci 15:3952–3969

Zanos P, Moaddel R, Morris PJ, Georgiou P, Fischell J, Elmer GI, Alkondon M, Yuan P, Pribut HJ, Singh NS, Dossou KS, Fang Y, Huang XP, Mayo CL, Wainer IW, Albuquerque EX, Thompson SM, Thomas CJ, Zarate CA Jr, Gould TD (2016) NMDAR inhibition-independent antidepressant actions of ketamine metabolites. Nature 533:481–486

Zarate CA Jr, Singh JB, Carlson PJ, Brutsche NE, Ameli R, Luckenbaugh DA, Charney DS, Manji HK (2006) A randomized trial of an N-methyl-D-aspartate antagonist in treatment-resistant major depression. Arch Gen Psychiatry 63:856–864

Zhang XM, Yan XY, Zhang B, Yang Q, Ye M, Cao W, Qiang WB, Zhu LJ, Du YL, Xu XX, Wang JS, Xu F, Lu W, Qiu S, Yang W, Luo JH (2015) Activity-induced synaptic delivery of the

GluN2A-containing NMDA receptor is dependent on endoplasmic reticulum chaperone Bip and involved in fear memory. Cell Res 25:818–836
Zhong J, Russell SL, Pritchett DB, Molinoff PB, Williams K (1994) Expression of mRNAs encoding subunits of the *N*-methyl-D-aspartate receptor in cultured cortical neurons. Mol Pharmacol 45:846–853
Zhou Q, Sheng M (2013) NMDA receptors in nervous system diseases. Neuropharmacology 74:69–75
Zhou C, Douglas JE, Kumar NN, Shu S, Bayliss DA, Chen X (2013) Forebrain HCN1 channels contribute to hypnotic actions of ketamine. Anesthesiology 118:785–795

Synaptic Modulation in the Effect of Ketamine

Daisuke Okada

Abstract Our central nervous system constantly instructs movements, generates recognition, and calculates value and emotion. Constant adaptation to the environment may be achieved by the integration of emotion/reward and recognition. Some diseases of mind may come from failure in making integration adaptive to environment. Neuronal circuits are established by the experience-dependent synaptic plasticity of glutamatergic neurons. Many of such main routes are equipped with the recurrent inhibition by GABAergic interneurons and regulated by monoaminergic modulation representing the reward and emotion. Major depression disease is considered as a state of dysfunction of these circuits which is here referred to as maladaptation. Studies of antidepressant mechanisms of ketamine include how stress induces changes in the original circuit, and how ketamine achieves the recovery of its function. In this chapter both lines of studies are discussed from the view points of the excitatory synapse hypothesis of major depression, and mechanisms of transient and persistent synaptic plasticity including reconsolidation and extinction learning.

Keywords Ketamine · Enantiomers · Hydroxynorketamine · NMDA receptor · Neuronal circuit · Synaptic plasticity · Maladaptation · Associativity · BDNF

1 Introduction

This chapter briefly introduces molecular and cellular mechanisms of persistent memory to discuss their possible involvement in major depression disease (MDD), and antidepressant actions of ketamine and its related compounds including metabolites.

D. Okada (✉)
Molecular and Cellular Neuroscience, Department of Medical Sciences, Graduate School of Kitasato University, Sagamihara, Japan

Department of Biochemistry, School of Medicine, Kitasato University, Sagamihara, Japan
e-mail: dada@med.kitasato-u.ac.jp

K. Hashimoto et al. (eds.), *Ketamine*,
https://doi.org/10.1007/978-981-15-2902-3_5

1.1 Ketamine as a Unique Antidepressant

Ketamine has been known as a dissociative anesthetic and an analgesic (Tyler et al. 2017). As a psychotomimetic drug, ketamine elicits both positive and negative symptoms which bear close resemblance to schizophrenia (Krystal et al. 1994). Now ketamine is attracting significant attention as a unique antidepressant in many aspects (Berman et al. 2000). First, ketamine shows rapid and persistent effects as an antidepressant. Remission appears in MDD patients within 2 h after intravenous injection of ketamine and lasts for a week (Zarate et al. 2006). Second, ketamine has wider therapeutic effects on mood disorders. It alleviates refractory MDD which does not respond to conventional antidepressants such as SSRIs (Zarate et al. 2006). It also reduces suicidal thought (Wilkinson et al. 2018) and is effective also on bipolar disease (Diazgranados et al. 2010). Extensive efforts have been made to understand the mechanisms underlying the antidepressant effect of ketamine, which will lead us to invent a better antidepressant without aversive effects of ketamine. The rapid and persistent effect implies that ketamine attacks the center of the neuronal mechanisms by which depressive symptoms are produced. Studies further found that ketamine enhanced synaptic plasticity which is associated with expression of brain-derived neurotrophic factor (BDNF), another compound known to suppress depression (Shirayama et al. 2002). Recently, it was found that (2*R*,6*R*)-hydroxynorketamine (HNK), one of the ketamine metabolites, has the antidepressant effect like ketamine (Zanos et al. 2016, 2017; Suzuki et al. 2017). (R)-HNK at antidepressant-relevant concentrations did not affect N-methyl-D-aspartate (NMDA) receptor currents (Suzuki et al. 2017; Zanos et al. 2017), indicating that ketamine interacts with unknown target molecules other than NMDA receptors to show antidepressant effects.

1.2 Antidepressant Actions of Ketamine and Circuit Behaviors

Ketamine is known as a noncompetitive antagonist of the NMDA-type of glutamate receptors (Anis et al. 1983). NMDA receptor plays pivotal roles in synaptic plasticity (Collingridge et al. 1983), which is the central mechanism underlying experience-dependent establishment and maintenance of neuronal circuits in the brain (Neves et al. 2008). Because ketamine blocks NMDA receptor, it is likely that ketamine blocks plastic changes in glutamatergic synapses. However, it is reported that ketamine facilitates long-term potentiation (LTP), a typical mode of synaptic plasticity associated with sustained enhancement of synaptic transmission. Ketamine enhances responses of α-amino-3-hydroxy-5-methylisoxazole-4-propionate (AMPA) receptors, another and the major glutamate receptors that generate excitatory postsynaptic potentials. In addition, ketamine induces the expression of BDNF, which happens in persistent LTP (Autry et al. 2011). Furthermore, ketamine administration enhanced gamma wave oscillation, a phasic brain activity enhanced during higher brain functions (Nugent et al. 2018). These observations collectively indicate that ketamine elicits the activation of neuronal connection through LTP-like enhancement of glutamatergic synapses.

1.3 Reconsolidation and Excitatory Synapse Hypothesis

Synaptic plasticity has been studied as a cellular basis of activity-dependent formation and maintenance of neuronal circuit (Bliss and Lømo 1973). Among higher functions of the brain, memory can be tested through behavioral tests in human and laboratory animals. LTP is currently considered as the cellular basis of establishment and maintenance of memory.

Episodic memory is memories of everyday life events, including context such as space, time order, persons and objects, and novelty, and generated in the hippocampus. Simultaneously, emotional values are calculated in another limbic structure, the amygdale. Episodic memory is acquired by experience, and its retention periods are related to the strength of the emotional impact or attention during experience. The strong memory is "consolidated" after the acquisition and can be recalled long after the experience (consolidation of the memory) (Wang and Morris 2010).

When the second experience results in a similar consequence to the first, the memory is enhanced. This reconsolidation is a mechanism of maintenance and even reinforcement of the original memory after recall (Wang and Morris 2010). On the other hand, when the second experience results in a different consequence from the first, recall of the memory of the first experience is inhibited. Sometimes a new memory is built for the second experience (Phelps et al. 2004). This is extinction learning. Aversive memories can be reduced by facilitating extinction learning or inhibition of reconsolidation (Lee et al. 2006; Shiller et al. 2010).

The excitatory synapse hypothesis of MDD (Thompson et al. 2015) considers the possibility that glutamatergic synapses in the corticomesolimbic reward circuit are affected by chronic stress. This system includes dopaminergic and serotonergic modulation of glutamate outputs, in accordance with the effects of many antidepressant drugs. However, this idea by itself does not explain the long-lasting effect of ketamine, which is a characteristic advantage of ketamine. Memory mechanisms based on the synaptic plasticity and reconsolidation may explain MDD as learned maladaptation of the neuronal circuits representing emotion and memory.

Plasticity-dependent maladaptive circuit hypothesis (Fig. 1) may suggest mechanisms of ketamine action. When experience is modest, synaptic plasticity makes the neuronal circuit functions adapted to the environments (Fig. 1a), while successive stressful experience modulates monoaminergic transmission and generates maladaptation of neuronal circuit (Fig. 1b). Ketamine blocks the activity of the malfunctioning circuit by interacting with NMDA receptor or other target molecules. Also, ketamine is considered to elicit a sustained recovery of the healthy behaviors through a revival of the original, adapted circuit or the activation of substituting circuits (Fig. 1c). Antidepressants having affinity to glutamatergic circuits such as ketamine may inhibit reconsolidation and reduces reinforcement effects of continuous stress. Alternatively, they may facilitate extinction learning. The reward circuit may be the major target, but it includes multiple regions such as the medial prefrontal cortex, the hippocampus, the nucleus accumbens, the ventral tegmental area, the amygdala, and the lateral habenula. Furthermore,multiple targets for the antidepressant action of ketamine (Zanos et al. 2016) are suggested. To evaluate the hypothesis, region and target molecules of ketamine action should be specified.

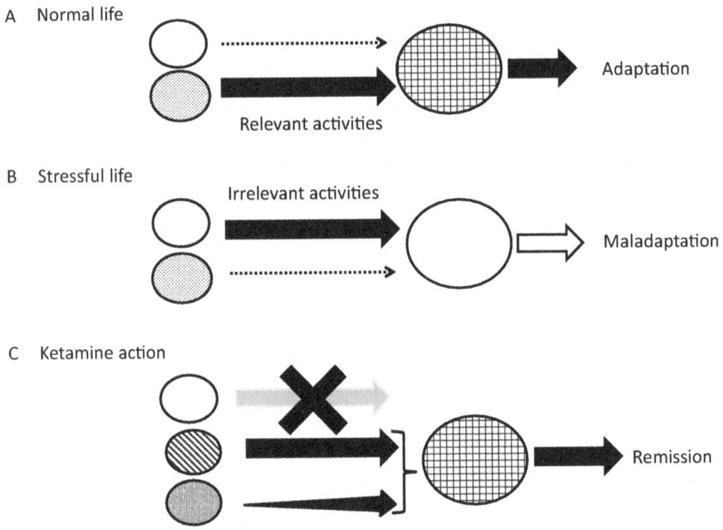

Fig. 1 Plasticity-dependent maladaptive circuit hypothesis. (a) Sensory inputs from environments and internal conditions in ordinary everyday life (shaded circle at the left) usually establish (thick arrow) adaptive neuronal circuits (meshed circle in center) through persistent plasticity. Unusual inputs (open circle at the left) usually fail to establish its circuit (dotted arrow). (b) Stress may give different types of inputs to the neurons which cause transient changes in behaviors. Prolonged or recursive stressful life generates persistent plasticity probably through activation of the monoamine modulatory inputs of the reward and emotional systems (thick arrow). The neuronal circuit specifically responding to robust but rare inputs will be established; however, it is not adaptive to usual environment (open arrow). In MDD, normal adaptive circuit may be masked (dotted arrow). (c) Ketamine is supposed to have two effects on the maladaptive circuits. First, sensory integration involving irrelevant signals, which is dominant in MDD, is inhibited by effects of ketamine (a cross on thin arrow) by means of transient disruption of transmission or inhibition of reconsolidation of the irrelevant circuit. Second, the adaptive circuit is restored by BDNF-dependent LTP. Two ways of restoration are considered. Case 1 is substitution (thick arrow from hatched circle that is representative of alternative adaptive conditions, at the left), which is achieved by establishment of a distinct link between sensory and integration systems by consolidation. Case 2 is revival (growing arrow from shaded circle), which is achieved by reactivation of the original adaptive circuit by extinction learning. Both possibilities suggest that the ketamine action should be long lasting to complete the remission. To make appropriate LTP, the context recognition of the patient should be supported, for example by association of occupational therapy

2 Persistent Synaptic Plasticity, Consolidation and Reconsolidation of Memory

Here, I summarize molecular and cellular mechanisms of synaptic plasticity, which may be critically involved in antidepressant effects of ketamine.

2.1 *Synaptic Plasticity and Neuronal Circuit Formation*

Brain functions are executed by neuronal circuits in which neurons are connected by synapses. Presynaptic release of neurotransmitters depolarizes (EPSP) or hyperpolarizes (IPSP) the postsynaptic cell by binding to postsynaptic receptors. Amplitude of postsynaptic potentials is variable (graded). When the depolarization at the initial segment of the postsynaptic axon exceeds the threshold, action potentials are evoked (firing), by which the excitation of the postsynaptic cell is transferred to the next cells. Thus, a neuronal circuit is composed of neurons linked to each other by dominant synapses which frequently contribute to firing of the postsynaptic cells.

Observation of unitary transmission revealed that principal neurons in the central nervous system (CNS) often fail to evoke EPSP, suggesting that the transmission has low efficiency. Furthermore, among many EPSP-evoking events, a limited number of EPSP can contribute to generate action potentials. How a synapse becomes dominant? Efficacy of synaptic transmission is regulated by several mechanisms, among which use-dependent plasticity is a mechanism that allows the circuit to behave in a manner adaptive to the environment. Hebb (1949) proposed a hypothesis in which functional neuronal circuits are established by a rule of "win for the used." Namely, a synapse may be dominant and survive if its activity frequently contributes to postsynaptic firing. Each of the principal neurons in the brain, such as pyramidal cells in the cerebral cortices, has tens of thousands of synapses receiving distinct presynaptic fibers carrying distinct activities from a variety of precedent neurons. Dominant synapse selectively carries one of such activities that frequently happen in the given environment. Synaptic plasticity is a mechanism by which dominant synapses are made in an activity-dependent manner. Plastic modulation of projection neurons is likely to contribute in making functional connections between distant brain areas.

Another possible role of plasticity may be the association. One neuron receives multiple input signals at different synapses which are mutually independent, while a neuron has only one output axon. Therefore, a single neuron can participate in multiple lines of signals through different synapses. This may cause the association of distinct events. Association of memory may be important to construct the abstract and collective concepts, our intelligence, and even our personality. Although these considerations are not supported by experimental evidence, we can imagine that if such an association accidentally takes place among irrelevant experiences, the brain recognizes the experience, or environment, with irrelevant value. In this chapter, maladaptation of neuronal circuit means misassociation between signals in the reward system which leads to dysfunction of the system.

2.2 Synaptic Plasticity

LTP was found by Bliss and Lømo (1973) in perforant path/granule cell synapses in the rabbit dentate gyrus and seemed to conform to Hebb's hypothesis. The LTP establishment includes presynaptic, postsynaptic, and perisynaptic mechanisms, and a dominant mechanism governing the LTP seems to differ between synapses. LTP in the hippocampal CA1 pyramidal cell/Schaffer collateral synapses as well as the dentate gyrus granule cell/perforant path synapses occurs predominantly under the postsynaptic mechanisms. LTD (long-term depression, a plastic decrease of synaptic transmission) in Purkinje cell/parallel fiber synapse occurs also by the postsynaptic mechanism. In contrast, LTP in the hippocampal CA3 pyramidal cell/mossy fiber synapse occurs predominantly under the presynaptic mechanism. It is noted that synaptic efficacy can be altered by changes in every one of the residents of synapses, such as transmitter availability, exocytosis kinetics, synaptic cleft geometry, availability and sensitivity of postsynaptic receptors, and excitability. Research with superresolution microscopy revealed that presynaptic and postsynaptic machinery are allocated so that transmission occurs effectively. For instance, presynaptic centers of exocytosis and postsynaptic receptor scaffolding proteins are localized face-to-face in the synaptic membranes (Sakamoto et al. 2018).

Although LTP was found in vivo experiments, its molecular mechanisms were revealed mainly by electrophysiological and cell biological experiments using acute brain slices. Transient, short-term LTP was evoked in CA1 of hippocampal slices by electrical stimulation of presynaptic axons at 100 Hz for 1 s (for example Okada et al. 1989). Spike-timing dependent plasticity in the cerebral cortex was usually evoked by brief depolarization of postsynaptic cell together with weak activation of presynaptic axon (Markram et al. 1997). In CA1 and dentate synapses, these stimuli activate NMDA receptor, which resulted in an increase in the number of AMPA-type glutamate receptors (functional plasticity), mainly GluA1 subunits, expressed in postsynaptic plasma membranes (surface expression) (Shi et al. 2001). It is associated with spine head enlargement by actin filament polymerization (morphological plasticity) (Matsuzaki et al. 2004). LTD in this synapse is a persistent decrease in EPSP amplitude and spine shrinkage (Zhou et al. 2004). LTD is a distinct type of plasticity and should be discriminated from depotentiation which is a counterreaction of LTP (Malenka and Bear 2004). These changes are regulated by the kinase cascade reactions, including calcium-calmodulin-dependent kinases (CaMKs) and mitogen-activated protein kinases (MAPKs) (Thomas and Huganir 2004). The effects of protein phosphorylation are reversible due to dephosphorylation and ubiquitin-dependent degradation of the protein (Jouvenceau et al. 2003). Thus, the expression of short-term LTP depends on phosphorylation of preexisting synaptic proteins. It is reversible and referred to as transient LTP. Expression mechanisms of LTP/LTD differ among synapses. For example, the reduction in spine head diameter is not observed in LTD of cerebellar Purkinje cell/parallel fiber synapses (Sdrulla and Linden 2010).

2.3 *Molecular and Cellular Mechanisms of the Transient LTP*

Establishment of transient LTP has three important characteristics: cooperativity, input-specificity, and associativity. Studies of the molecular mechanisms of LTP revealed relevance of these features in memory.

1. Cooperativity: Strong input is required for LTP establishment.

   Strong input is supplied by repetitive activity at high frequency of the presynaptic cell or robust depolarization of postsynaptic cell by activity of a second presynaptic axon. Sensory system always receives signals from the outer world and conditions of the body. Therefore, if LTP arises from tiny excitation of neurons, our sensation may be filled up with noise. Synaptic plasticity with cooperativity guarantees the stable and contrasted sensation, which may be advantageous for adaptive behaviors.

   The magnitude of Ca^{2+} influx (peak and duration) is an essential determinant for direction of the plastic change (Mizuno et al. 2001), suggesting that the cooperativity of LTP is regulated by intracellular Ca^{2+} concentrations (Malenka et al. 1988). The spike timing-dependent plasticity protocol revealed that higher intracellular Ca^{2+} concentrations lead to LTP, while LTD appears in lower Ca^{2+} concentrations (Dan and Poo 2004).
2. Input specificity: LTP is established exclusively in the synapse which received inputs sufficient for the cooperativity.

   LTP in one synapse is known to activate surrounding synapses by diffusion of activated molecules (Hedrick and Yasuda 2017), which alters the sensitivity of the neighboring synapse for triggering plasticity, a phenomenon known as metaplasticity. However, in general, crosstalk between the original synapse and unrelated axons having distinct information is prevented by the input specificity mechanisms, so that the cooperative activity evokes LTP only in the selected synapses. The coincidence detection mechanisms such as the depolarization-induced release of Mg^{2+}-blockade of NMDA receptor is known as the input-specificity mechanisms of short-term plasticity (Tsien 2000). This mechanism evokes robust intracellular Ca^{2+} influx only when both pre- and postsynaptic cells are depolarized in CA1 pyramidal cells (Nowak et al. 1984). The postsynaptic Ca^{2+} increase evokes LTP which is confined in these synapses.

   Optogenetic experiments with genetically modified mice showed that the memory of the learned tasks was recalled by reactivation of synapses which had been activated during learning (training in experiments) (Liu et al. 2014), whereas specific ablation of the synapses that had been activated during training selectively destroyed the memory of the learned task (Hayashi-Takagi et al. 2015). These results strongly support the synapse-dependent formation of neuronal circuit according to Hebb's hypothesis.

Input-specific LTP supports establishment of the input-relevant circuit. Failure in the input-specificity mechanism may cause dysfunction of the circuit. It is possible that the malfunction of input-specificity causes irrelevant coupling of unrelated experiences. This may cause erroneous or inadequate recognition. In MDD, confused association of thoughts or emotion may cause irregular thoughts characteristic of the patients.

3. Associativity: When robust depolarization and a weak input come to a neuron simultaneously, LTP appears in both synapses. The LTP evoked in weakly stimulated synapse in association with strong heterosynaptic activity is called associative LTP.

Associative plasticity is a cellular model of association memory. Behavioral experiments indicate that associative memory is formed by training using combinatorial stimuli, a learning protocol known as classical conditioning. Consider that the "stimulus A" evokes the "behavior A", while the "stimulus B" evokes the "behavior B". Importantly, the "stimulus A" does not evoke the "behavior B" by itself. When both "stimulus A" and "stimulus B" are given simultaneously, the animal sometimes learns the connection between two stimuli and then the "stimulus A" becomes to evoke the "behavior B". In psychological protocol of the conditioned learning, "stimulus A" is the conditioned stimulus, and "stimulus B" is the unconditioned stimulus.

To make an associative memory, it is assumed that the synapse commonly activated by both stimuli A and B should be generated in a neuron participating in the circuit. Such common synapses are referred to as behavioral tag (Ballarini et al. 2009; Nomoto et al. 2016). Recent study demonstrated that the association of false memory can be artificially established, suggesting that the behavioral tag associated two behaviours (Ohkawa et al. 2015). The behavioral tagging mechanism suggests a hypothesis for MDD, in which the maladaptive neuronal circuit is formed by irrelevant selection of association pair. Continuous experience of extreme life events or hard environments may affect emotional circuits which in turn enhance plasticity to form the behavioral tag in irrelevant synapse leading to maladaptive behaviors.

2.4 *Consolidation of Episodic Memory and Its Molecular Components*

Synaptic plasticity has the possibility to be a major mechanism to establish maladaptive neuronal circuit that can be related to MDD. Although transient LTP or LTD is reversible, the MDD pathology is long lasting, suggesting that ketamine is likely to affect the persistent type of plasticity.

Long-lasting memory can be recalled by looking back long time after the initial experience. According to psychology of the memory, long-lasting memories are consolidated as a stable memory. In the words of neuronal cell biology, memory consolidation is establishment of a stable neuronal circuit, in other words, synaptic connections between neurons become stably robust. These neurons are reactivated synchronously by a cue input of a component of the episode, which is recall of the memory. Consolidation of memory is considered to be supported by persistent synaptic plasticity which requires the following four cellular events (Reymann and Frey 2007):

1. Transient plasticity: In the persistent LTP, both functional and morphological changes evoked in the transient LTP become persistent. Synaptic functions and structures are changed by transient LTP, which may work as the preparation for the persistent reforming of synapse by newly synthesized proteins (Inoue et al. 2007).
2. *de novo* protein synthesis: Persistent LTP requires induction of gene expression and is blocked by inhibitors of transcription and protein synthesis. Long-term memory requires reinforcement inputs such as dopamine transmission which activates cyclic AMP-responsive element (CRE)-dependent induction of transcription (Silva et al. 1998). Although the precise nature of reinforcement inputs is not clear, persistent LTP enhances CRE binding protein phosphorylation in the nucleus and new protein synthesis takes place in the soma and dendrites. Animal experiments reported that expression of about 800 genes was changed with various time lags according to genes (Ryan et al. 2012). Since newly synthesized protein is the critical for the persistent plasticity, cooperativity of persistent LTP is assigned to protein synthesis-evoking stimulus such as dopaminergic input.
3. Allocation mechanism of newly synthesized proteins: Newly synthesized proteins function only in the activated synapses. Following two mechanisms are known to manage this. These mechanisms support input-specificity of the persistent plasticity through localization of the newly synthesized proteins.

   "Synaptic tagging and capture" hypothesis assumes that the transient plasticity activates the "synaptic tag" in the activated synapses. A synaptic tag enables the newly synthesized synaptic proteins to function in the synapse to cause persistent plasticity (Frey and Morris 1997). This hypothesis assumes that the newly synthesized proteins in soma are unspecifically transported along most of dendrites, while these proteins function specifically in the synapse where the transient plasticity took place (Sajikumar and Frey 2004). Molecular mechanism of synaptic tagging for Homer-1a protein was unraveled (Okada et al. 2009). Protein trafficking is not free between dendritic spine and dendrite (Bloodgood and Sabatini 2005). Homer-1a protein is synthesized in soma, unspecifically transported in most dendrites, and then enters spines only where synaptic NMDA receptor is activated. Thus, the synaptic tag of Homer-1a protein is the activity-dependent regulation of spine entry of the protein. Synaptic tags are possibly variable depending on proteins.

Local synthesis is another allocation mechanism that supplies proteins such as Ark and GluA1, by translation in dendrites, at the gate of spines or within spines (Everwine et al. 2001). Local synthesis occurs in a mTORC-dependent manner, using preexisting translational machinery such as initiation and elongation factors and ribosomes (Schuman 1999). The dendritic transport of the mRNA of particular genes is the important component of local synthesis (Wang et al. 2009). AMPA receptors synthesized in the dendrites are incorporated in the synaptic plasma membranes (Ju et al. 2004). Experiments suggested that both local synthesis and synaptic tagging are necessary for persistent associative LTP and they are likely not mutually exclusive. Ryan et al. (2012) reported that changes in expression of over 800 genes were detected during 5 and 24 h after induction of persistent LTP in vivo, which shows different sets of genes are activated at different time, probably due to the cascade activation of transcription factors. Therefore, synaptic tagging and local synthesis may regulate allocation of distinct sets of gene products within the distinct time windows (Okada and Inokuchi 2015).

4. Reconstruction of the synapse: Persistent LTP enhances AMPA receptor-dependent current, spine head diameter, and postsynaptic density (PSD) size in a sustained manner. Newly synthesized proteins are localized in the activated synapses, and believed to reconstruct the molecular architecture of the synapse so that enhanced numbers of synaptic molecules may function stably. To reconstruct synaptic molecular complex, the existing complex is destroyed at first. This scrap-and-build process reforms the enlarged synapse, thereby achieving stable enhancement of synaptic function and morphology. For example, Homer-1c makes scaffolding networks in the PSD (Hayashi et al. 2009). Homer-1a is an immediate early gene activated in persistent LTP in the hippocampal CA1 and induces breakdown of the Homer-1c-dependent protein complex in the PSD, and this breakdown is necessary for building the enlarged Homer-1c complex (Inoue et al. 2007).

2.5 Recall and Reconsolidation of Episodic Memory

Neuronal circuit that supports an episodic memory is established by transient plasticity at the moment of experience. It is stabilized (consolidated) by persistent plasticity, which is associated with stable enlargement of postsynaptic protein complex. However, the memory recall by presentation of cues that remind the original experience is known to destroy the synaptic molecular complexes by ubiquitin-proteasome systems (Lee et al. 2008). Nevertheless, a second or a third presentation of the episode often succeeds recalling the memory, suggesting that the recall cue both destroys the consolidated synaptic complex and builds it again to hold the memory. In our everyday life, one can notice that a second memory often differs from the first impression. Actually, the molecular structure supporting the first memory is destroyed by recall and rebuilt anew by reconsolidation mechanism. Reconsolidation of the hippocampal contextual memory involves activity in the cortex. Reconsolidation of the emotional memory obtained by conditioning in the amygdale is also regulated by the cortex. Reconsolidation is a memory updating process which enhances the original memory when a second experience is very similar to the first (Wang and Morris 2010).

When a second experience differs from the first in its emotion or reward, then the cortical activity masks the original memory circuit in the amygdala to prevent its reactivation. The memory circuit continue to exist but cannot be activated, while a new memory is generated according to the second consequence. Thus, this phenomenon is called extinction learning. The original memory often relapse depending on the magnitude of cortical inhibition. Extinction mechanisms involve many distinct brain structures such as hippocampus and prefrontal cortex, as well as nucleus accumbens for reward memory and amygdale for emotional memory. For example, infralimbic medial prefrontal cortex (ILmPFC) integrates memories and when a second experience does not involve emotional cue, ILmPFC inhibits amygdale output, resulting in extinction learning of amygdale memory (Quirk and Mueller 2008).

2.6 *Modulation of Gamma Oscillation by Ketamine*

Gamma wave is oscillatory activity at 30–90 Hz and attracts attention due to its relationship between higher brain functions such as memory. Changes in gamma wave oscillation are considered as the brain-wide activity representing depression. On the other hand, ketamine enhance both gamma power and AMPA receptor currents, suggesting the antidepressant effect of ketamine comes from LTP of the gamma oscillation producing pyramidal cells. Studies with mathematical models suggested that gamma wave requires the recurrent micro-circuit consisted of a pyramidal neuron which is recurrently inhibited by Parvalbumine-expressing basket cells (PVBCs) (Bartos et al. 2007). These microcircuits make pyramidal cells fire synchronously. The calcium-permeable GluA2-absent AMPA receptors are the dominant source of Ca^{2+} influx triggering neurotransmitter release from PVBC (Goldberg et al. 2003). Providing contribution of NMDA receptor is small, inhibition of NMDA receptors by ketamine in this circuit may not affect the gamma wave power (Gonzalez-Burgos and Lewis 2012). Thus, relationship between gamma wave and ketamine action is still not clear.

3 Possible Target Molecules of Ketamine

3.1 NMDA Receptors

Dizocilpin (MK801), phencyclidine, and ketamine are non-competitive antagonist of NMDA receptor channels. They are open channel blockers which enter the receptor's open pore to occlude it. MK801 is specific to NMDA receptor and irreversibly bind to it, while ketamine binding is low affinity (McDonald et al. 1991). Ketamine has dual effects toward excitatory and inhibitory synapses. NMDA receptor pore in neurons at resting potential is occupied by extracellular Mg^{2+} ions. Therefore, ketamine inhibits NMDA receptor only when it is open. It is likely that inhibition of NMDA receptors in excitatory postsynapses suppresses LTP.

NMDA receptors are tetrameric and composed of 2 subunits of GluN1 and 2 subunits of GluN2 or GluN3. Glutamate binds to GluN2 subunits, whereas GluN1 and GluN3 bind glycine and D-serine. GluN2 receptors have 4 isoforms, GluN2A, B, C and D. The affinity of ketamine for these receptors is nearly equal (Paoletti and Neyton 2007). However, NMDA receptors containing GluN2C or 2D elicit inward current by unitary glutamate transmission because their channels show weaker Mg^{2+}-blockade (Larsen et al. 2014). GluN2B/D-containing channels are expressed in the axonal terminals of the cerebellar stellate cells, which are inhibitory. GluN2B/D mediates spontaneous activity of these interneurons at resting potential, which is required for the long-term release of GABA (Dubois et al. 2016). Blockade of NMDA receptors with weak Mg^{2+}-blockade in the axon terminals of inhibitory neurons suppresses the output from resting neurons, causing disinhibition of the circuit (Akgül and McBain 2016). Neuronal circuits composed of a main output of glutamatergic principal neuron, recurrent GABAergic inhibitory neuron, and monoaminergic modulatory inputs are often seen in the brain as important role players for recognition and diseases (Carlen et al. 2012; Nakazawa et al. 2012; Ren et al. 2016). Experiments with knockout mice showed that antidepressant effect of (*R*)-ketamine required GluN2D receptor (Ide et al. 2017).

In the lateral habenula, NMDA receptor is involved in generation of bursting action potentials in inhibitory neurons such as PVBCs. This bursting is enhanced in MDD-like conditions (Yang et al. 2018a). Bursting activity of inhibitory PVBC strongly inhibits projection neurons and the downstream monoamine systems. Ketamine inhibits NMDA receptors, which blocks the bursting. It is noted that Kir4.1 channel expression is enhanced in MDD-like state (Cui et al. 2018). Increased expression of potassium channels causes depolarization of neurons by decreasing extracellular potassium concentrations, which is the cause of the NMDA receptor-dependent bursting of the neuron. After all, it seems that the mechanism of enhanced expression of Kir4.1 is critically linked to MDD pathology, but ketamine seems to inhibit MDD only transiently. Persistent antidepressant effect of ketamine may suggest involvement of plastic change in this circuit.

3.2 D-Serine and Other NMDA Receptor Ligands

Binding of co-agonists, D-serine, or glycine to the GluN1 subunit is prerequisite of glutamate-dependent activation of the receptor. Action of these co-agonists depends on the local extracellular concentrations. D-serine is synthesized by serine racemase (SR) in neurons. D-amino acid oxidase in astrocyte is responsible for its breakdown to pyruvate and hydrogen peroxide. D-serine is excreted to extracellular space through Asc-1 transporter. These molecules control extracellular D-serine concentrations. According to Papouin et al. (2012), NMDA receptors localized in pre- and post-synaptic membranes bind D-serine, while those localized in soma bind glycine. Decrease in extracellular D-serine attenuated LTP (Henneberger et al. 2010). D-serine interacts with other molecules such as cerebellar GluD2 and α7-nicotinic acetylcholine receptor (α7nAChR). α7nAChR increases SR expression through mTORC1 activation, while (*S*)-ketamine inhibits α7nAChR and reduces NMDA receptor activity (Paul et al. 2014).

Kynurenine is a tryptophan metabolite, and its derivative 4-chlorokynurenine is an NMDA receptor antagonist competitive with D-serine which is reported to show antidepressant effect (Zanos et al. 2015). An antibody against the glycine site, B6B21, has the partial agonist activity. Glyxins are peptides having amino acid sequences resembling the supervariable region of the antibody. Glyxin13 was delivered across the blood–brain barrier (BBB) and enhanced some forms of learning performance (Moskal et al. 2005).

MK801 and AZD6765 are open channel blockers of the NMDA receptor channel. MK801 showed specific and intense binding to NMDA receptors, while those of AZD6765 are weak. These drugs showed only transient antidepressant effects (Zarate et al. 2013; Maeng et al. 2008). By comparing this observation and the fact that ketamine, known to act at the same locus as MK801, shows persistent antidepressant effect, it is suggested that transient and long-lasting effects of ketamine may be resulted from multiple actions to different targets.

Ro25-6981 is a selective inhibitor of the GluN2B containing NMDA receptors. Its effects were not consistent between studies. Maeng et al. (2008) reported transient antidepressant effects, whereas Li et al. (2011) observed long-lasting effects although different behavioral tests were employed to estimate the antidepressant effects in these reports.

3.3 Enantiomers and Metabolites of Ketamine

Enantiomers and enantiomeric metabolites of ketamine are studied on the potency for use as antidepressant and results differ among studies (Ide and Ikeda 2018). Ketamine has high BBB permeability and its bioavailability is high when administered via intramuscular injection, but it is low via oral administration, suggesting that metabolic degradation in the liver is dominant (Ebert et al. 1997).

N-demethylation of ketamine gives norketamine, which in turn 6-hydroxylated to give 6-HNK. 6-HNK is also synthesized by 6-hydroxylation first followed by N-demethylation. Hydroxylation occurs also at 4- and 5-positions, but the major metabolite is 6-HNK. Ketamine has a racemic center at the second carbon, therefore consists of (*S*)- and (*R*)-enantiomers. Hydroxylation does not change the chirality; thus, (2*S*,6*S*)- and (2*S*,6*R*)-HNK are derived from (*S*)-ketamine and (2*R*,6*S*)- and (2*R*,6*R*)-HNK are derived from (*R*)-ketamine. Among these *cis–trans* isomers, the *cis* types are dominant; therefore, the major metabolites of (*S*)- and (*R*)-ketamine are (2*S*,6*S*)- and (2*R*,6*R*)-HNK, respectively (Zanos et al. 2016).

These metabolic reactions are carried out by cytochrome P450 (CYP). According to Tyler et al. (2017), N-demethylation involves CYP2A6, 2B4, 2C19, 3A4, and 3A5, while 6-hydroxylation of norketamine involves CYP2A6, 2B6, and 3A5. However, identification of the CYP isoforms involved in ketamine metabolism varied between studies, though CYP2A6 and 2B6 are reported as the common candidates. CYP has many isoforms and the variety differs among species. For example, the CYP2B6 gene is found in human, while it is absent in mice which have CYP2B1 as the highest homology. CYPs are expressed mainly in the liver, the adrenal gland, and the gonad glands. Some CYPs involved in estrogen synthesis (Hojo et al. 2004) and nicotine metabolism (Miskys et al. 2000) are found to function in the hippocampal neurons. Although HNK and ketamine readily cross BBB, ketamine was not metabolized in the rat brain homogenate, suggesting the major contribution of the liver CYPs in HNK production (Moaddel et al. 2016).

Some metabolites of ketamine are known to have antidepressant effects as well as psychomimetic effects. Ketamine and norketamine are NMDA receptor blockers and work as anesthetics and analgesics. (*S*)-compounds are 5 times more intense as an NMDA receptor blocker than (*R*)-compounds. *K*i values for each compounds for replacement of [^{3}H] MK-801 bound in rat brain membranes were: (*S*)-ketamine 0.3 μM vs. (*R*)-ketamine 1.4 μM, and (*S*)-norketamine 1.7 μM vs. (*R*)-norketamine 13 μM (Ebert et al. 1997). (*S*)-norketamine is more rapidly produced than (*R*)-norketamine.

Both enantiomers of ketamine are reported to have an antidepressant effect. Reus et al. (2015) reported that antidepressant effect of (*S*)-ketamine against the depression-like behaviors in the forced swimming test, the splash test, and the open-field test in adult rat received maternal deprivation. Zhang et al. (2014) reported, using juvenile mice received neonatal dexamethason exposure, acute antidepressant effect of both enantiomers of ketamine on the tail suspension test, the forced swimming test, and the sucrose preference test. Interestingly, they observed that (*R*)-ketamine selectively showed reduction in the immobility in the tail suspension and forced swimming tests 1 week after injection, suggesting (*R*)-ketamine has long-lasting antidepressant effects. These results suggest the possibility that the acute and long-lasting effects of ketamine were created by distinct enantiomers.

(2*R*,6*R*)-HNK, at lower concentrations than inhibiting NMDA receptor, was reported to show rapid and persistent antidepressant effects (Zanos et al. 2016, 2017). (2*R*,6*R*)-HNK raised the BDNF expression and the AMPA current enhancement in the hippocampus, suggesting that persistent LTP is established at least a part. AMPA receptor currents are not enhanced by other NMDA receptor blockers. These results suggest that (2*R*,6*R*)-HNK elicits antidepressant effects through unknown target molecules other than NMDA receptors. Interestingly, Yang et al. (2018b) reported that (*S*)-norketamine showed more intense antidepressant effect against depressive behaviors induced by inflammation or chronic social defeat stress than (*R*)-norketamine. They also reported that (*R*)-ketamine but not (2*R*,6*R*)-HNK, had significant antidepressant effect.

These studies do not reach a complete agreement. Effects of these compounds should be tested using a common battery of behavioral test protocols. It is also important to note that these compounds have a distinct contribution on the acute and persistent antidepressant effects.

3.4 Raft and BDNF

Orser et al. (1997) suggested that ketamine affects NMDA receptors through two distinct mechanisms. One is filling in the open channel pore from the extracellular space and the other is having access to the receptor's allosteric site facing plasma membrane. The latter implies a possibility of ketamine action in the membrane. Wray et al. (2018) reported that in C6 glioma cells, ketamine treatment released Gαs proteins from lipid raft. Gαs activated adenylyl cyclase in the cell out of the raft and facilitated cyclic AMP-dependent events such as CRE-dependent BDNF expression. This response was not evoked by other NMDA receptor ligands, but (2*R*,6*R*)-HNK evoked similar response. They also showed that Gαs release from raft may involve tubulin acetylation. HDAC6 inhibition and knock out elicited antidepressant effects (Singh et al. 2018). This work suggests three independent possibilities. First, ketamine action does not necessarily require actions mediated by neurotransmitter receptors. Second, acetylation–deacetylation balance in the glial cells may be associated with cell biology of the depressed brains. Third, the ketamine targeted on the glia affects the nearby neurons through action of glia-derived BDNF.

Antidepressant effect of ketamine involves the synthesis and action of BDNF (Bramham and Messaoudi 2005). Observation of mTORC1 (Autry et al. 2011) and MAPK (Yang et al. 2018b) activation by ketamine support this idea. Loci of BDNF synthesis and action are ambiguous in these reports, while neuronal action of glia-derived BDNF is suggested (Wray et al. 2018). BDNF expression is the key step of persistent plasticity. CRE-dependent translation is triggered by binding of phosphorylated CREB protein onto the CRE (Malburg and Blendy 2005). Many protein kinases are reported to phosphorylate CREB, such as cyclic AMP-dependent protein kinase, CaMKII, GSK3β, Akt, and MAPKs. BDNF is released from the production cell by exocytosis. Plasma BDNF concentrations are low in MDD. Monoamine neurotransmitters facilitate LTP and BDNF production through CRE-dependent translation activation. BDNF production may be an important mechanism of antidepressant effects of SSRI. BDNF action is mediated by TrkB receptors, which in turn activates cascade reactions of kinases composed of MEK (a MAPK kinase), ERK (a

MAPK), and Akt. The MAPK cascade facilitates major expression mechanisms of LTP such as GluA1 trafficking and CRE-dependent transcription activation. Akt phosphorylation triggers mTORC1-dependent protein synthesis, which is involved in local protein synthesis.

References

Akgül G, McBain CJ (2016) Diverse roles for ionotropic glutamate receptors on inhibitory interneurons in developing and adult brain. J Physiol 594:5471–5490

Anis NA, Berry SC, Burton NR et al (1983) The dissociative anaesthetics, ketamine and phencyclidine, selectively reduce excitation of central mammalian neurons by N-methyl-D-aspartate. Br J Pharmacol 79:565–575

Autry AE, Adachi M, Nosyreva E et al (2011) NMDA receptor blockade at rest triggers rapid behavioural antidepressant responses. Nature 475:91–95

Ballarini F, Moncada D, Martinez MC et al (2009) Behavioral tagging is a general mechanism of long-term memory formation. Proc Natl Acad Sci U S A 106:14599–14604

Bartos M, Vida I, Jonas P (2007) Synaptic mechanisms of synchronized gamma oscillations in inhibitory interneuron networks. Nat Rev Neurosci 8:45–56

Berman RM, Cappiello A, Anand A et al (2000) Antidepressant effects of ketamine in depressed patients. Biol Psychiatry 47:351–354

Bliss TVP, Lømo T (1973) Long-lasting potentiation of synaptic transmission in the dentate gyrus of the anaesthetized rabbit. J Physiol (Lond) 232:331–356

Bloodgood BL, Sabatini BL (2005) Neuronal activity regulates diffusion across the neck of dendritic spines. Science 310:866–869

Bramham CR, Messaoudi E (2005) BDNF function in adult synaptic plasticity: the synaptic consolidation hypothesis. Prog Neurobiol 76:99–125

Carlen M, Meletis K, Siegle JH et al (2012) A critical role for NMDA receptors in parvalbumin interneurons for bamma rhythm induction and behavior. Mol Psychiatry 17:537–548

Collingridge GL, Kehl SJ, McLennan H (1983) Excitatory amino acids in synaptic transmission in the Schaffer collateral-commissural pathway of the rat hippocampus. J Physiol 334:33–46

Cui Y, Yang Y, Ni Z et al (2018) Astroglial Kir4.1 in the lateral habenula drives neuronal bursts in depression. Nature 554:323–327

Dan Y, Poo MM (2004) Spike timing-dependent plasticity of neural circuits. Neuron 44:23–30

Diazgranados N, Ibrahim L, Brutsche NE et al (2010) A randomized add-on trial of an N-methyl-D-aspartate antagonist in treatment-resistant bipolar depression. Arch Gen Psychiatry 67:793–802

Dubois CJ, Lachamp PM, Sun L et al (2016) Presynaptic GluN2D receptors detect glutamate spillover and regulate cerebellar GABA release. J Neurophysiol 115:271–285

Ebert B, Mikkelsen S, Thorkildsen C et al (1997) Norketamine, the main metabolite of ketamine, is a non-competitive NMDA receptor antagonist in the rat cortex and spinal cord. Eur J Pharmacol 333:99–104

Everwine J, Miyashiro K, Kacharmina JE et al (2001) Local translation of classes of mRNAs that are targeted to neuronal dendrites. Proc Natl Acad Sci U S A 98:7080–7085

Frey U, Morris RGM (1997) Synaptic tagging and long-term potentiation. Nature 385:533–536

Goldberg JH, Yuste R, Tamas G (2003) Ca2+ imaging of mouse neocortical interneurone dendrites: contribution of Ca^{2+}-permeable AMPA and NMDA receptors to subthreshold. J Physiol 551:67–78

Gonzalez-Burgos G, Lewis DA (2012) NMDA receptor hypofunction, Parvalbumin-positive neurons, and cortical gamma oscillations in schizophrenia. Schizophr Bull 38:950–957

Hayashi MK, Tang C, Verpelli C et al (2009) The postsynaptic density proteins Homer and Shank form a polymeric network structure. Cell 137:159–171

Hayashi-Takagi A, Yagishita S, Nakamura M et al (2015) Labelling and optical erasure of synaptic memory traces in the motor cortex. Nature 525:333–338
Hebb DO (1949) The organization of behavior. Wiley, New York
Hedrick NG, Yasuda R (2017) Regulation of Rho GTPase proteins during spine structural plasticity for the control of local dendritic plasticity. Curr Opin Neurobiol 45:193–201
Henneberger C, Papouin T, Oliet S et al (2010) Long-term potentiation depends on release of D-serine from astrocytes. Nature 463:232–236
Hojo Y, Hattori T, Enami T et al (2004) Adult male rat hippocampus synthesizes estradiol from pregnenolone by cytochromes P45017α and P450 aromatase localized in neurons. Proc Natl Acad Sci U S A 101:865–870
Ide S, Ikeda K (2018) Mechanisms of the antidepressant effects of ketamine enantiomers and their metabolites. Biol Psychiatry 84:551–552
Ide S, Ikekubo Y, Mishina M et al (2017) Role of NMDA receptor GluN2D subunit in the antidepressant effects of enantiomers of ketamine. J Pharmacol Sci 135:138–140
Inoue Y, Udo H, Inokuchi K et al (2007) Homer1a regulates the activity-induced remodeling of synaptic structures in cultured hippocampal neurons. Neuroscience 150:841–852
Jouvenceau A, Billard JM, Haditsch U et al (2003) Different phosphates-dependent mechanisms mediate long-term depression and depotentiation of long-term potentiation in mouse hippocampal CA1 area. Eur J Neurosci 18:1279–1285
Ju W, Morishita W, Tsui J et al (2004) Activity-dependent regulation of dendritic synthesis and trafficking of AMPA receptors. Nat Neurosci 7:244–253
Krystal JH, Laurence P, Seibyl JP et al (1994) Subanesthetic effects of the noncompetitive NMDA antagonist, ketamine in humans. Arch Gen Psychiatry 51:199–214
Larsen R, Corlew RJ, Henson MA et al (2014) NR3A-containing NMDARs promote neurotransmitter release and spike timing-dependent plasticity. Nat Neurosci 14:338–344
Lee JLC, Milton AL, Everitt BJ (2006) Reconsolidation and extinction of conditioned fear: inhibition and potentiation. J Neurosci 26:10051–10056
Lee SH, Choi JH, Lee N, Lee HR, Kim JI, Yu NK, Choi SL, Lee SH, Kim H, Kaang BK (2008) Synaptic protein degradation underlies destabilization of retrieved fear memory. Science 319:1253–1256
Li N, Liu RJ, Dwyer JM et al (2011) Glutamate N-methyl-D-aspartate receptor antagonists rapidly reverse behavioral and synaptic deficits caused by chronic stress exposure. Biol Psychiatry 69:754–761
Liu X, Ramirez S, Redondo RL et al (2014) Identification and manipulation of memory engram cells. Cold Spring Harb Symp Quant Biol 79:59–65
Maeng S, Zarate CA, Du J et al (2008) Cellular mechanisms underlying the antidepressant effects of ketamine: role of α-amino-3-hydroxy-5-methylisoxazole- 4-propionic acid receptors. Biol Psychiatry 63:349–352
Malburg JE, Blendy JA (2005) Antidepressant action: to the nucleus and beyond. Trends Pharmacol Sci 26:631–638
Malenka RC, Bear MF (2004) LTP and LTD: an embarrassment of riches. Neuron 44:5–21
Malenka RC, Kauer JA, Zucker RS et al (1988) Postsynaptic calcium is sufficient for potentiation of hippocampal synaptic transmission. Science 242:81–84
Markram H, Lübke J, Frotscher M et al (1997) Regulation of synaptic efficacy by coincidence of postsynaptic APs and EPSPs. Science 275:213–215
Matsuzaki M, Honkura N, Ellis-Davies GCR et al (2004) Structural basis of long-term potentiation in single dendritic spines. Nature 429:761–766
McDonald JF, Bartlett MC, Pahapill MP et al (1991) Actions of ketamine, phencyclidine and MK801 on NMDA receptor currents in cultured mouse hippocampal neurones. J Physiol 432:483–508
Miskys S, Hoffman E, Tyndale RF (2000) Regional and cellular induction of nicotine-metabolizing CYP2B1 in rat brain by chronic nicorine treatment. Biochem Pharmacol 59:1501–1511
Mizuno T, Kanazawa I, Sakurai M (2001) Differential induction of LTP and LTD is not determined solely by instantaneous calcium concentration: an essential involvement of a temporal factor. Eur J Neurosci 14:701–708

Moaddel R, Sanghvi M, Dossou KSS et al (2016) The distribution and clearance of (2S,6S)-hydroxynorketamine, an active ketamine metabolite, in Wistar rats. Pharmacol Res Perspect 3(4):e00157. https://doi.org/10.1002/prp2.157

Moskal JR, Kuo AG, Weiss C et al (2005) GLYX-13: a monoclonal antibody- derived peptide that acts as an N-methyl-D-aspartate receptor modulator. Neuropharmacology 49:1077–1087

Nakazawa K, Zsiros V, Jiang Z et al (2012) GABAergic interneuron origin of schizophrenia pathophysiology. Neuropharmacology 62:1574–1583

Neves G, Cooke SF, Bliss TVP (2008) Synaptic plasticity, memory and the hippocampus: a neural network approach to causality. Nat Rev Neurosci 9:65–75

Nomoto M, Ohkawa N, Nichizono H et al (2016) Cellular tagging as a neural network mechanism for behavioural tagging. Nat Commun 7:12319. https://doi.org/10.1038/ncomms12319

Nowak L, Bregestovski P, Ascher P et al (1984) Magnesium gates glutamate-activated channels in mouse central neurones. Nature 307:462–465

Nugent AC, Ballard ED, Gould TD et al (2018) Ketamine has distinct electrophysiological and behavioral effects in depressed and healthy subjects. Mol Psychiatry 24(7):1040–1052. https://doi.org/10.1038/s41380-018-0028-2

Ohkawa N, Saitoh Y, Suzuki A et al (2015) Artificial association of pre-stored information to generate a qualitatively new memory. Cell Rep 11:261–269

Okada D, Inokuchi K (2015) Activity-dependent protein transport as a synaptic tag. In: Sajikumar SR (ed) Synaptic tagging and capture. Springer, New York, pp 79–98

Okada D, Yamagishi S, Sugiyama H (1989) Differential effects of phospholipase inhibitors in long-term potentiation in the rat hippocampal mossy fiber synapses and Schaffer/commissural synapses. Neurosci Lett 100:141–146

Okada D, Ozawa F, Inokuchi K (2009) Input-specific spine entry of soma-derived Vesl-1S protein conforms to synaptic tagging. Science 324:904–909

Orser BA, Pennefather PS, MacDonald JF (1997) Multiple mechanisms of ketamine blockade of N-methyl-D-aspartate receptors. Anesthesiology 86:903–917

Paoletti P, Neyton J (2007) NMDA receptor subunits: function and pharmacology. Curr Opin Pharmacol 7:39–47

Papouin T, Ladepeche L, Ruel J et al (2012) Synaptic and extrasynaptic NMDA receptors are gated by different endogenous coagonists. Cell 150:633–646

Paul RK, Singh NS, Khadeer M et al (2014) (R,S)-ketamine metabolites (R,S)-norketamine and (2S, 6S)-hydroxynorketamine increase the mammalian target of rapamycin function. Anesthesiology 121:149–159

Phelps EA, Delgado MR, Nearing KI et al (2004) Extinction learning in humans: Role of the amygdale and vmPFC. Neuron 43:897–905

Quirk GJ, Mueller D (2008) Neural mechanisms of extinction learning and retrieval. Neuropsychopharmacol Rev 33:56–72

Ren Z, Pribiag H, Jefferson SJ et al (2016) Bidirectional homeostatic regulation of a depression-related brain state by gamma-aminobutyric acidergic deficits and ketamine treatment. Biol Psychiatry 80:457–468

Reus GZ, Carlessi AS, Titus SE et al (2015) A single dose of S-ketamine induces long-term antidepressant effects and decreases oxidative stress in adulthood rats following maternal deprivation. Dev Neurobiol 75:1268–1281

Reymann KG, Frey JU (2007) The late maintenance of hippocampal LTP: requirements, phases, 'synaptic tagging', 'late-associativity' and implications. Neuropharmacology 52:24–40

Ryan MM, Ryan B, Kyrke-Smith M et al (2012) Temporal profiling of gene networks associated with the late phase of long-term potentiation in vivo. PLoS One 7:e40538. https://doi.org/10.1371/journal.pone.0040538

Sajikumar S, Frey JU (2004) Late-associativity, synaptic tagging, and the role of dopamine during LTP and LTD. Neurobiol Learn Mem 82:12–25

Sakamoto H, Ariyoshi T, Kimpara N et al (2018) Synaptic weight set by Munc13-1 supramolecular assemblies. Nat Neurosci 21:41–49

Schuman EM (1999) mRNA trafficking and local protein synthesis at the synapse. Neuron 23:645–648
Sdrulla AD, Linden DJ (2010) Double dissociation between long-term depression and dendritic spine morphology in cerebellar Purkinge cells. Nat Neurosci 10:546–548
Shi AH, Hayashi Y, Esteban JA et al (2001) Subunit-specific rules governing AMPA receptor trafficking to synapses in hippocampal pyramidal neurons. Cell 105:331–343
Shiller D, Monfis M-H, Raio CM et al (2010) Preventing the return of fear in humans using reconsolidation update mechanisms. Nature 463:49–53
Shirayama Y, Chen ACH, Nakagawa S et al (2002) Brain-derived neurotrophic factor produces antidepressant effects in behavioral models of depression. J Neurosci 22:3251–3261
Silva AJ, Kogan JH, Frankland PW et al (1998) CREB and memory. Annu Rev Neurosci 21:2127–2148
Singh H, Wray N, Schappi JM et al (2018) Disruption of lipid-raft localized Gαs/tubulin complexes by antidepressants: a unique feature of HDAC6 inhibitors, SSRI and tricyclic compounds. Neuropsychopharmacology 43:1481–1491
Suzuki K, Nosyreva E, Hunt KW et al (2017) Effects of a ketamine metabolite on synaptic NMDAR function. Nature 546:E1–E3. https://doi.org/10.1038/nature22084
Thomas GM, Huganir RL (2004) MAPK cascade signaling and synaptic plasticity. Nat Rev Nurosci 5:173–183
Thompson SM, Kallarackal AJ, Kvarta MD et al (2015) An excitatory synapse hypothesis of depression. Trends Neurosci 38:279–294
Tsien JZ (2000) Linking Hebb's coincidence-detection to memory. Cirr Opin Neurobiol 10:266–273
Tyler MW, Yourish HB, Ionescu DF et al (2017) Classics in chemical neuroscience: ketamine. ACS Chem Nerosci 8:1122–1134
Wang SH, Morris RGM (2010) Hippocampal-neocortical interactions in memory formation, consolidation, and reconsolidation. Annu Rev Psychol 361:22.1–22.31
Wang DO, Kim SM, Zhao Y et al (2009) Synapse- and stimulus-specific local translation during long-term neuronal plasticity. Science 324:1536–1540
Wilkinson ST, Ballard ED, Bloch MH et al (2018) The effect of a single dose of intravenous ketamine on suicidal ideation: a systematic review and individual participant data meta-analysis. Am J Psychiatry 175:150–158
Wray NH, Schappi JM, Singh H et al (2018) NMDAR-independent, cAMP-dependent antidepressant actions of ketamine. Mol Psychiatry 24(12):1833–1843. https://doi.org/10.1038/s41380-018-0083-8
Yang Y, Cui Y, Sang K et al (2018a) Ketamine blocks bursting in the lateral habenula to rapidly relieve depression. Nature 554:317–322
Yang C, Kobayashi S, Nakao K et al (2018b) AMPA receptor activation-independent antidepressant actions of ketamine metabolite (S)-Norketamine. Biol Psychiatry 84:591–600
Zanos P, Piantadosi SC, Wu HQ et al (2015) The prodrug 4-Chlorokynurenine causes ketamine-like antidepressant effects, but not side effects, by NMDA/GlycineB-site inhibition. J Pharmacol Exp Ther 355:76–85
Zanos P, Moaddel R, Morris P et al (2016) NMDAR inhibition-independent antidepressant actions of ketamine metabolites. Nature 533:481–486
Zanos P, Moaddel R, Morris PJ et al (2017) Zanos et al. reply. Nature 546:E4–E5. https://doi.org/10.1038/nature22085
Zarate CA, Jaskaran BS, Carlson PJ et al (2006) A randomized trial of an N-methyl-D-aspartate antagonist in treatment-resistant major depression. Arch Gen Psychiatry 63:856–864
Zarate CA, Mathews D, Ibrahim L et al (2013) A randomized trial of a low-trapping nonselective N-methyl-D-aspartate channel blocker in major depression. Biol Psychiatry 74:257–264
Zhang JC, Li SX, Hashimoto K (2014) R(-)-ketamine shows greater potency and longer lasting antidepressant effects than S(+)-ketamine. Pharmacol Biochem Behav 116:137–141
Zhou Q, Homma K, Poo MM (2004) Shrinkage of dendritic spines associated with long-term depression of hippocampal synapses. Neuron 44:749–757

Behavioral Pharmacology of Ketamine: An Overview of Preclinical Studies

Naoyuki Hironaka

Abstract The behavioral pharmacological effects of ketamine in animals have been reviewed. Ketamine does not cause remarkable behavioral changes at doses effective for the treatment of depressive behavior in animal models. However, a transient increase in spontaneous motor activity has been reported sometimes. Ketamine is effective for the recovery of behavior caused by chronic mild stress, social defeat stress, and so on. Several mechanisms underlying this effect have been proposed, but a conclusive answer has still not been presented. (R)-Ketamine is more potent and has longer-lasting antidepressant effects than the (S)-isomer. Ketamine produces diverse behavioral changes other than antidepressant effects, such as blunting fear, cognitive impairments, and social withdrawal. Behavioral changes related to schizophrenia-like symptoms such as impairment of prepulse inhibition of the acoustic startle response and impairment of latent inhibition have been reported. However, relatively high doses seem to be necessary to yield these effects. Several lines of preclinical evidence have also shown the reinforcing and rewarding properties of ketamine, which are relevant to the abuse liability. To overcome these side effects and to optimize its clinical efficacy as a rapid-onset antidepressant for treatment-resistant patients, detailed preclinical studies targeting proper dosing regimen for clinical applications will be important.

Keywords Preclinical studies · Gross behavioral observation · Animal models of depression · Emotion · Cognition · Social interaction · Abuse liability

Abbreviations

im	Intramuscular
ip	Intraperitoneal
iv	Intravenous
sc	Subcutaneous

N. Hironaka (✉)
Department of Pharmacology, LSI Medience Corp, Tokyo, Japan

K. Hashimoto et al. (eds.), *Ketamine*,
https://doi.org/10.1007/978-981-15-2902-3_6

1 Introduction

Ketamine, a dissociative anesthetic agent, was developed in the early 1960s as a substitute for phencyclidine (PCP). The term dissociative means that it differentially acts on subcortical and cortical regions in the brain. Indeed, it is known that ketamine alters the functional connectivity of the human brain (Joules et al. 2015). The mechanisms of action of ketamine are complex. However, it is well known that the major target site of ketamine is the *N*-methyl-D-aspartate (NMDA) receptor. Ketamine blocks the opening of the channel and thereby reduces mean channel open time and at the same time decreases the frequency of channel opening via an allosteric mechanism (Orser et al. 1997). Further, it is also known that ketamine inhibits monoamine transporters and therefore potentiates monoaminergic neurotransmission (Nishimura et al. 1998).

Ketamine has been widely used as an anesthetic, especially for short diagnostic and surgical procedures since it does not cause muscle relaxation. It is a standard anesthetic in veterinary medicine. However, the nonmedical use of ketamine has increasingly been recognized as an abuse-related problem. In Europe, concomitant use of ketamine with 3,4-methylenedioxymethampehtamine (MDMA) for recreational purposes has become a concern. In 2003, a WHO expert committee on drug dependence warned that a critical review regarding ketamine is needed (WHO technical report series, 915, 2003: 15). In Japan, ketamine was designated as a narcotic drug in the Narcotics and Psychotropics Control Act in 2007.

Recently, the antidepressant action of ketamine has attracted attention. One of the major drawbacks of traditional antidepressants is the slow onset of their therapeutic efficacy. Both preclinical and clinical studies have shown that a single administration of ketamine is effective for the treatment of depression (see review, Grady et al. 2018; Zanos and Gould 2018). Although ketamine is expected to be a promising antidepressant drug, we must deliberately evaluate other behavioral effects. Ketamine produces diverse behavioral effects on emotion and cognition. It also causes schizophrenia-like behavioral symptoms. Moreover, ketamine has reinforcing and rewarding properties, which lead to abuse.

In this chapter, I will review the behavioral pharmacological effects of ketamine in animals. This review focuses on (1) its effects on gross behaviors; (2) the effects related to its therapeutic efficacy as an antidepressant; and (3) its effects on emotion, cognition, and those related to abuse liability, which might cause adverse reactions in clinical settings.

2 General Behavioral Effects

Subanesthetic doses of ketamine do not cause remarkable behavioral changes in animals. We conducted gross behavioral observations on the effects of ketamine in rats by means of the functional observational battery (FOB) used in our laboratory.

The battery covers a wide range of behavioral changes such as tremor, convulsion, respiration, reactivity to handling, exophthalmos, salivation, piloerection, pupil size, rearing, posture, urination, defecation, and sensory reactivity to stimuli. According to our observation, ketamine at 3 mg/kg ip increased the number of rearing from 5 to 60 min after administration. However, this sign was not clear at 10 mg/kg. No other remarkable changes were observed.

Ketamine at relatively high doses seems to increase spontaneous motor activity. For example, ketamine at 10–100 mg/kg ip temporarily increased locomotion, sniffing, swaying, and falling. These signs were similar to those of PCP and different from stimulant drugs (Koek et al. 1987). In mice, ketamine increased motor activity at 30 mg/kg ip (Hayase et al. 2006; Lin et al. 2016). However, at 100 mg/kg ip motor activity was decreased after a temporary increase (Hayase et al. 2006). In rats, ketamine at 25 mg/kg ip increased activity and stereotypy for 30 min after administration (Razoux et al. 2007).

The effects of ketamine on steady-state and ongoing behavior have been studied by means of schedule-controlled behavior for food reinforcement established by operant conditioning. Ketamine at doses up to 10 mg/kg ip had no effect on the nose-poking behavior of mice. Doses higher than 30 mg/kg decreased the response rate maintained by a fixed ratio (FR) 30-s schedule, in which every 30 responses were reinforced and a relatively high baseline response was maintained. On the contrary, in responses maintained by the fixed interval (FI) 300-s schedule, in which the first response occurred 300 s after the previous reinforcement was reinforced and a relatively low baseline response was maintained. Ketamine at 30–100 mg/kg ip increased the response rate but decreased the response rate at the highest concentration of 180 mg/kg (Wenger and Dews 1976). The effect of ketamine on schedule-controlled behavior varies depending on the preadministration baseline rate of the response. This effect is called rate dependence and is common to various kinds of psychoactive drugs. The rate-dependent effect of ketamine has also been demonstrated in S-ketamine at doses above 15 mg/kg ip in rats under the FI 300 s schedule (Meliska et al. 1980). However, they showed that R-ketamine at doses above 30 mg/kg ip did not increase the response rate but only decreased it.

3 Behavioral Effects Related to Antidepressant Therapeutic Efficacy

3.1 Overview

In 1988, Reynolds and Miller reported that tricyclic antidepressants, such as desmethylimipramine and imipramine, inhibited [^{3}H]-MK801 binding to the NMDA receptor. The IC_{50} value of desmethylimipramine was 7.4 μM and that of imipramine was 22.5 μM (Reynolds and Miller 1988). This in vitro finding was confirmed in vivo by analysis of the protective action of ketamine against lethality by a large dose (up to 80 mg/kg ip) of NMDA in mice (Leander 1989). The fact that antidepressant drugs have antagonistic activity on NMDA receptors led to the glutamate hypothesis of depression (see review Sanacora et al. 2012).

In accordance with this hypothesis, the clinical efficacy of ketamine on depression patients has been reported in the early 2000s. For example, Berman et al. (2000) showed that a single administration of ketamine at 0.5 mg/kg iv in patients with major depression was effective for improving symptoms as measured by the Hamilton Depression Rating Scale. Although the number of patients in this study was small, the effect of ketamine was clear compared with the placebo control. Preclinical studies on the antidepressant effects of ketamine using animal models have been accelerated after that.

3.2 *Findings with Various Models of Depression*

In an early study, ketamine at 10 mg/kg ip recovered suppression of sucrose preference and novelty-food seeking after chronic mild stress in rats (Li et al. 2011). In this study, it was also shown that ketamine recovered the expression levels of synapsin-1 and PSD95 and excitatory postsynaptic currents in the prefrontal cortex.

Lipopolysaccharide (LPS) induces cytokines such as interleukin-1β (IL-1β) and tumor necrosis factor-α (TNF-α) and it is known that these proinflammatory cytokines are elevated in depression patients. Thus, administrating LPS is used to induce depression-like behavioral changes in animals (Zhu et al. 2010). In the LPS-induced depression model in mice, ketamine at 6 mg/kg ip recovered sucrose preference and reduced immobility time in forced swimming test. This effect was thought to be mediated by antagonism of the NMDA receptor because the effect of ketamine was blocked by pretreatment with the α-amino-3-hydroxy-5-methylisoxazole-4-propionic acid (AMPA) receptor antagonist 2,3-dihydroxy-6-nitro-7-sulfamoylbenzo(f)quinoxaline-2,3-dione (Walker et al. 2013).

Social isolation is another method to induce depression-like behavioral changes in animals (Wallace et al. 2009). Isolated housing for 7–8 weeks caused a decrease in sucrose preference and prolonged immobility time in the forced swimming test in male rats, whereas a decrease in sucrose preference was not evident in females (Sarkar and Kabbaj 2016). In this study, ketamine at a relatively low dose (2.5 and 5 mg/kg ip) improved these symptoms. It was also found that isolated housing decreased spine density and levels of synapsin-1, PSD95, and GluR1 in the medial prefrontal cortex and these changes were also restored by ketamine.

Another important aspect of depression is a loss of cognitive flexibility (Castaneda et al. 2008). Subjecting rats to cold stress (4 °C) daily for 6 h for 14 consecutive days induced an impairment of cognitive flexibility, as revealed by impairment of reversal learning in the attentional set-shifting experiment (Fig. 1). A single administration of ketamine at 10 mg/kg ip improved performance in this test (Patton et al. 2017). This effect may be mediated by the Janus tyrosine kinases and signal transduction activators of transcription proteins (JAK/STAT) signaling pathway.

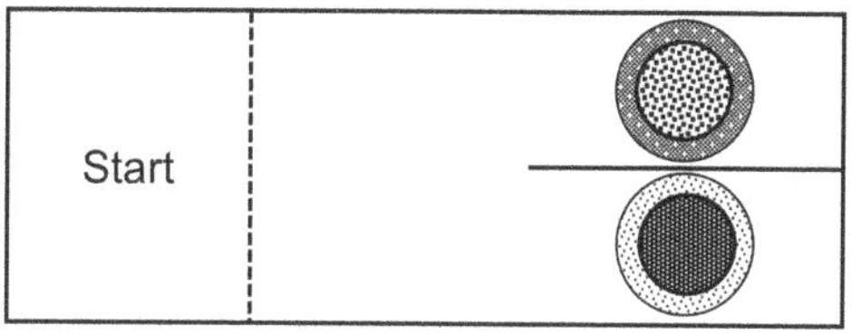

Fig. 1 Reversal learning in attentional set-shifting apparatus in rats. Food-restricted rats were placed in the start box (left) and allowed to select one of the two pods (right). Each pod contained a different digging medium and a different odor applied to the rim of the pods. In the original discrimination training, rats had to associate one stimulus dimension (e.g., odor) with food. The other dimension (e.g., digging medium) was an irrelevant distractor. In the reversal learning, the cue/reward association was switched. (This figure is originally drawn based on Lapiz and Morilak (2006))

3.3 *Quest for Mechanism*

Because the mechanisms of action of ketamine related to its antidepressant effect will be discussed in detail in another chapter, I will only cover several select topics relevant to its behavioral effects.

The rapid onset of the antidepressant effects of ketamine has attracted attention. In an earlier study, it was demonstrated that blockade of the NMDA receptors by ketamine deactivates the eukaryotic elongation factor 2 (eEF2) kinase and reduces phosphorylation of eEF2. This induces rapid synthesis of brain-derived neurotrophic factor (BDNF), resulting in an antidepressant effect (Autry et al. 2011).

The NMDA receptor in the cortex is a heteromultimeric complex and contains two GluN1 subunits and two GluN2 subunits. Of the GluN2 subunits, GluN2A and GluN2B subunits are dominant. The target site of the antidepressant effects of ketamine were studied using cortical neurons from mice with a selective deletion of GluN2B (Miller et al. 2014). In these mice, the effect of a single administration of ketamine at 50 mg/kg ip did not reduce immobility time in the tail suspension test, suggesting that this subunit is one of the major sites of action of the antidepressant effects of ketamine. GluN2B seems to play an important role in the manifestation of depressive signs because mice having a deletion of this subunit are insensitive to chronic corticosterone treatment, which usually causes enhanced signs of depression.

Although ketamine is an NMDA receptor antagonist, several biochemical studies have shown that ketamine increases glutamate neurotransmission in the prefrontal cortex. As for this seemingly paradoxical effect of ketamine, one study suggested that ketamine acts on the NMDA receptors in hippocampal GABAergic interneurons as an antagonist and this antagonism activates the glutamate system in the prefrontal cortex (Donegan and Lodge 2017).

However, mechanisms other than glutamate neuronal transmission can also take part in the antidepressant effects of ketamine. One candidate mechanism involves the secretion of proinflammatory cytokines as mentioned in the LPS model. In one study, chronic unpredictable stress caused upregulation of cytokines such as IL-1β,

IL-6, TNF-α, and indoleamine 2,3-dioxygenase and this upregulation was attenuated by ketamine at 10 mg/kg ip (Wang et al. 2015).

Recently, a relationship between the gut microbiota and depression has been noted. One study demonstrated that treatment with ketamine at 2.5 mg/kg ip for 5 days markedly amplified *Lactobacillus, Turicibacter*, and *Sarcina*. In contrast, the same treatment decreased Mucispirillum and Ruminococcus, which are considered to be opportunistic pathogens (Getachew et al. 2018).

3.4 For Better Clinical Applications

Since ketamine has stereoisomers, it is important to know which isomer is potent for the treatment of depression. At the preclinical level, it is well known that (R)-ketamine is more potent than (S)-ketamine (Yang et al. 2015). It is also known that (R)-ketamine is longer lasting than (S)-ketamine (Fukumoto et al. 2017). (R)-Ketamine is effective in a social defeat stress model and this effect seems to be related to the gut microbiota–brain axis (Yang et al. 2017). This effect is independent of the serotonin system because pretreatment with para-chlorophenylalanine, which depletes brain serotonin, did not affect the antidepressant effect of ketamine (Zhang et al. 2018).

The antidepressant effect of ketamine has been proven in preclinical animal studies. However, there are still questions regarding scenarios that favor the action of ketamine the most. A recent study using mice highlighted this point. Ketamine at 10 and 30 mg/kg ip showed a reduction in the immobility time in the forced swimming test only in mice exposed to chronic mild stress. In contrast, ketamine increased the immobility time in normal unstressed mice (Fitzgerald et al. 2019). The response to ketamine is quite different in model mice compared to normal mice. This finding might be important because the effect is different from traditional antidepressants which are also effective in normal animals in the forced swimming test.

4 Behavioral Effects Related to Putative Adverse Reactions

Ketamine is known to cause schizophrenia-like symptoms in humans. The effects of ketamine on emotion, cognition, and social behaviors have been mainly studied in relation to this effect. It is also known to have abuse liability. These effects may cause adverse reactions in clinical settings. In this section, I will introduce the preclinical findings related to the putative adverse reactions. For convenience, the major findings are concisely summarized in Table 1.

Table 1 Behavioral effects of ketamine related to putative adverse reactions

Domain	Behavior	Species	Dose and route	Major finding	Ref.
Emotion	Anxiety	rt	30 mg/kg ip × 5 d	No effect in EPM	Becker et al. (2003)
	Fear	rt	16 mg/kg, sc	Suppression of conditioned fear	Pietersen et al. (2007)
	Fear	rt	up to 15 mg/kg ip × 3 d	Enhancement of fear in EPM after stress	Juven-Wetzler et al. (2014)
	Fear	rt	10 mg/kg ip	Enhancement of conditioned fear with pre-reactivation	Honsberger et al. (2015)
	Fear	ms	2.5 mg/kg ip × 22 d	facilitation of fear extinction with training	Ju et al. (2017)
Cognition	Memory	rt	30 mg/kg ip	Impairment of spatial short-term memory	Moghaddam et al. (1997)
	Memory	rt	30 mg/kg ip	Impairment of spatial memory in Morris maze	Duan et al. (2013)
	Memory	rt	30 mg/kg ip	Impairment of object and location recognition	Lin et al. (2016)
	Decision making	rt	up to 20 mg/kg ip	Increase impulsive choice in low impulse individuals	Cottone et al. (2013)
Schizophrenic	PPI of ASR	rt	30 mg/kg ip × 5 d	No effect	Becker et al. (2003)
	PPI of ASR	ms	50 mg/kg ip	Impairment at the lowest sound level	Featherstone et al. (2013)
	PPI of ASR	rt	30 mg/kg ip	Impairment	Duan et al. (2013)
	PPI of ASR	rt	30 mg/kg ip	Impairment	Lin et al. (2016)
	Latent inhibition	rt	30 mg/kg ip × 5 d	Impairment	Becker et al. (2003)
	Latent inhibition	rt	25 mg/kg ip	Impairment	Razoux et al. (2007)
	Executive function	mk	up to 1 mg/kg, im	Impairment of task switching performance	Stoet and Snyder (2006)
	Social withdrawal	rt	30 mg/kg ip × 5 d	Reduction of nonaggressive social contact	Becker and Grecksch (2004)
	Social withdrawal	rt	30 mg/kg ip × 5 d	Reduction of social contact	Uribe et al. (2013)

(continued)

Table 1 (continued)

Domain	Behavior	Species	Dose and route	Major finding	Ref.
Abuse liability	SA	mk	0.1 mg/kg/inf, iv	Reinforcing	Winger et al. (1989)
	SA	mk	0.03–0.3 mg/kg/inf, iv	Reinforcing	Winger et al. (2002)
	SA	rt	0.5 mg/kg/inf, iv	Reinforcing	van der Kam et al. (2007)
	SA	rt	0.125–1 mg/kg/inf. iv	Reinforcing only in novel environment	De Luca and Badiani (2011)
	SA	rt	0.5 mg/kg/inf, iv	Reinforcing but priming is required	Venniro et al. (2015)
	SA	rt	0.1 mg/kg/inf, iv	Reinforcing in males and proesterus	Wright et al. (2017)
	SA	rt	0.5 mg/kg/inf, iv	Reinforcing in OB rats	
	SA	rt	0.5 mg/kg/inf, iv	Reinforcing	Caffino et al. (2018)
	CPP	ms	1–10 mg/kg ip	Rewarding	Suzuki et al. (1999)
	CPP	ms	3, 10 mg/kg ip	Rewarding	Suzuki et al. (2000)
	CPP	rt	2.5, 5 mg/kg ip	Not rewarding but sensitization of hyperlocomotion	Strong et al. (2017)
	Other	rt	3.2–10 mg/kg ip	Not potentiate ICSS	Hillhouse et al. (2014)

PPI prepulse inhibition, *ASR* acoustic startle response, *CPP* conditioned place preference, *SA* self-administration, *ra* rat, *ms* mouse, *mk* monkey, *ip* intraperitoneal, *sc* subcutaneous, *im* intramuscular, *iv* intravenous

4.1 Emotion

There seems to be no direct effect of ketamine on anxiety. Ketamine at 30 mg/kg ip for 5 days did not cause any notable behavioral changes in the elevated-plus maze test (Becker et al. 2003).

However, ambiguous findings have been reported with respect to the effects of ketamine on fear memory. The effects seems to be dependent on environmental manipulations In one study, ketamine at 16 mg/kg sc suppressed freezing due to fear conditioning in rats, and this effect was antagonized by clozapine but not by haloperidol (Pietersen et al. 2007). In contrast, in another study, ketamine at 0.5–15 mg/kg ip for 3 days did not ameliorate but increased freezing elicited by a predator (cat)-scent exposure in rats (Juven-Wetzler et al. 2014). In this study, it was also noted that treatment with ketamine seemed to increase anxiety tested in the elevated-plus maze and the acoustic startle response.

If fear memory is reactivated by exposing animals to a conditioned context, ketamine at 10 mg/kg ip potentiated retrieval of fear memory in rats (Honsberger et al. 2015). However, if ketamine treatment (0.625–2.5 mg/kg ip for 22 days) is combined with extinction training of fear memory in mice, it reduces freezing and anxiety. This effect was related to the normalization of DNA methylation of the BDNF gene in the hippocampus and medial prefrontal cortex (Ju et al. 2017).

4.2 Cognition

Ketamine seems to deteriorate memory at relatively high doses. Short-term memory was impaired in rats administered 30 mg/kg ip as revealed by the delayed spatial alternation task (Moghaddam et al. 1997). Spatial memory, as tested in the Morris water maze, was also impaired by ketamine at 30 mg/kg ip (Duan et al. 2013). This effect was concomitant with hippocampal depression of neural transmission and these effects were antagonized by the dopamine D1 antagonist SCH23390 or the AMPA receptor endocytosis interfering peptide Tat-GluR2_{3Y}. Short-term memory as tested by novel object or novel location recognition was also impaired by ketamine at 30 mg/kg ip and this effect was recovered by coadministration with *N,N*-dimethylglycine (Lin et al. 2016).

Other than memory, impulsivity is related to various psychiatric diseases such as addiction and depression. Delay discounting is a measure used to evaluate impulsivity. In this test, preference for obtaining an immediate small reward and a delayed large reward is compared. Preference for an immediate small reward is considered an index of impulsivity. In one study, ketamine at 2.5–20 mg/kg ip has been reported to increase impulsive choices in rats. However, this effect was evident only for low-impulsive individuals (Cottone et al. 2013).

4.3 Symptomatology Related to Schizophrenia

Several characteristic information process biases relevant to the symptoms of schizophrenia are known. One of them is a deficit of prepulse inhibition (PPI) of the acoustic startle response (ASR). Abrupt large sounds elicit the startle response. If a weaker sound is presented immediately before this large sound, the amplitude of the ASR decreases. This is the PPI of ASR and is thought to reflect sensory-motor gating function. The degree of PPI is reduced, i.e., a large ASR is observed despite presenting a prepulse in patients with schizophrenia (Kunugi et al. 2007).

Although an early study did not demonstrate an impairment of PPI of ASR by ketamine at 30 mg/kg ip for 5 days for 2 weeks in rats (Becker et al. 2003), a subsequent study reported an impairment of PPI in mice, but only at the lowest level of sound pressure (Featherstone et al. 2013). This effect seemed to be mediated by the Akt 1 gene because the effect of ketamine was not evident in Akt 1 partially deleted mice. Other studies using ketamine at 30 mg/kg ip in rats also showed an impairment of the PPI (Duan et al. 2013; Lin et al. 2016).

Another characteristic related to the symptomatology of schizophrenia is the impairment of latent inhibition. If a neutral stimulus is presented several times prior to Pavlovian conditioning, the acquisition of the conditioned response will be retarded. This phenomenon is known as latent inhibition (LI). It is known that LI is impaired in schizophrenia patients (Swerdlow et al. 1996). Ketamine at 25 or 30 mg/kg ip for 5 days impaired LI in rats (Becker et al. 2003). Impairment of LI was also observed even after a single administration of ketamine at 25 mg/kg ip in rats (Razoux et al. 2007).

Executive functions, which are known to be deteriorated in schizophrenic patients, are also impaired by ketamine. In a study using rhesus monkeys, ketamine at doses up to 1 mg/kg, im, impaired executive function tested by a task switching experiment (Fig. 2) (Stoet and Snyder 2006).

Social withdrawal, one of the negative symptoms of schizophrenia, is induced by ketamine. Repeated administration of ketamine at a relatively high dose (30 mg/kg ip) in rats decreased social interaction; however, this effect was only evident in the nonaggressive component of social interaction (Becker and Grecksch 2004). This effect of ketamine was antagonized by clozapine but not by haloperidol. A similar dosing regimen also reduced social interaction and this effect was reversed by memantine, a modulator of the NMDA receptor (Uribe et al. 2013).

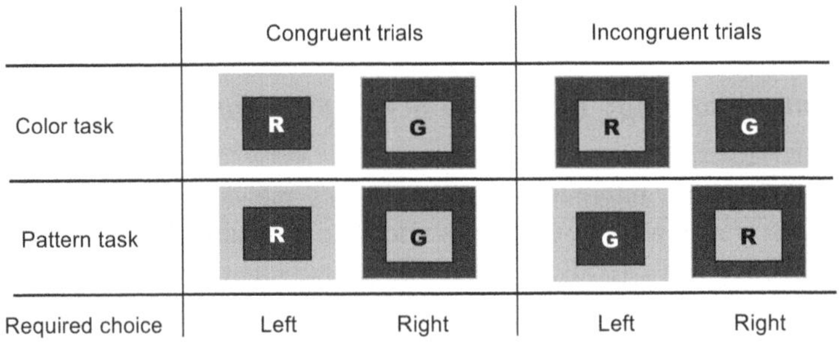

Fig. 2 Task switching experiment in monkeys. At the beginning of each trial, monkeys were instructed whether this was a color task or a pattern task by presenting a cue panel. In the color task, the monkey had to judge whether the color of the square was red or green by pressing two buttons (in this figure, red = left/green = right). In the pattern task, the monkey had to judge whether the square was brighter on the inside or on the outside (in this figure, outside = left/inside = right). In congruent trials, the correct choice side of both tasks was the same. In incongruent trials, however, the correct choice side of each task was different. (This figure is originally drawn based on Stoet and Snyder (2006))

4.4 Abuse Liability

Abuse liability, or in other words the dependence producing potential, of ketamine is discussed extensively in another chapter. In this section, I will summarize only the behavioral findings related to abuse liability.

One of the common features of an abused drug is its reinforcing property. The reinforcing property indicates the capability of a drug to act as a positive reinforcer of drug-taking behavior. In animals and also in humans, this property can be tested by drug self-administration experiments.

The reinforcing effects of ketamine have been reported in intravenous self-administration. Ketamine at 1.0 mg/kg/inf was self-administered in rhesus monkeys with a history of self-administration of cocaine or codeine (Winger et al. 1989). A subsequent self-administration study using a behavioral economics paradigm showed that ketamine had equal potency as a reinforcer to PCP (Winger et al. 2002). The reinforcing effect seems to be mediated by the glutamatergic system because the reinforcing effect of ketamine at 0.5 mg/kg/inf was decreased by the mGluR5 antagonist 2-methyl-6-(phenylethynyl) pyridine (MPEP) (van der Kam et al. 2007).

However, the reinforcing effect of ketamine was dependent on the environmental context, drug history, and sex. For example, self-administration of ketamine was evident only in rats transiently transferred from their home cages to the self-administration experiment cages. In habitual rats living in the self-administration experiment cage, the reinforcing effect of ketamine was not observed (De Luca and Badiani 2011). Moreover, the reinforcing effect of ketamine in drug naïve animals is difficult to show. Pretreatment with ketamine seems to be necessary to initiate self-administration (Venniro et al. 2015). In rats, intravenous ketamine at 0.1 mg/kg/inf was self-administered in males and proestrus females. Diestrus females did not show self-administration (Wright et al. 2017).

The rewarding effect of drugs is closely related to the reinforcing effect. The rewarding effect is tested by the conditioned place preference (CPP) experiment. In CPP, it is tested whether environmental stimuli associated with the drug effects induce an approaching behavior. In a classical study, ketamine at 10 to 30 mg/kg ip induced a dose-dependent CPP in mice and this effect was antagonized by the 5-HT_3 receptor antagonist ondansetron (Suzuki et al. 1999). The rewarding effect was confirmed in a subsequent study (3 and 10 mg/kg ip) (Suzuki et al. 2000). However, lower doses of ketamine (e.g., 2.5 or 5 mg/kg ip) in rats failed to induce CPP although ketamine causes behavioral sensitization to locomotor stimulating effects (Strong et al. 2017).

Several studies have been conducted to reveal the neurochemical mechanisms underlying the dependence-producing effect of ketamine. For example, the reinforcing property of ketamine was related to autophosphorylation of αCaMKII in the ventral striatum, medial prefrontal cortex, and hippocampus (Caffino et al. 2018). However, the precise mechanism of dependence to ketamine has not been elucidated to date. One of the most important candidates is its action on the brain reward system, the mesolimbic dopaminergic neurons originating from the ventral tegmentum area and projecting to the nucleus accumbens. Although ketamine increases glutamate release in the nucleus accumbens (Razoux et al. 2007), ketamine at 3.2–10 mg/kg ip did not potentiate the intracranial self-stimulation (ICSS) response, suggesting that ketamine has no direct effect on dopaminergic neurons in the brain reward system (Hillhouse et al. 2014).

5 Conclusion

A growing body of preclinical evidence has shown that (R)-ketamine is effective for the treatment of depression. Ketamine is expected to exert an antidepressant effect after a single administration in treatment-resistant patients. However, in preclinical experiments, a single administration of even traditional tricyclic antidepressants and SSRIs is effective. Thus, its rapid onset should be verified in human clinical studies. Although preclinical researchers have attempted to develop animal models of treatment-resistant depression by means of exposing animals to stress, further studies are needed to verify the efficacy in refractory depression. The precise mechanisms of the antidepressant effects are still under research and diverse mechanisms have been proposed. If treatment with ketamine is combined with behavioral intervention techniques, it is possible to ameliorate emotional fear responses, which are typically seen in PTSD. Although psychotomimetic effects and abuse liability may be cautioned as a source of putative adverse reactions in clinical settings, relatively high doses seem to be related to these effects. An appropriate dosing regimen for the treatment of depression should be considered based on the animal model studies.

Acknowledgment The author would like to thank Ms. Kikuyo Nakaoka, Department of Pharmacology, Kumamoto Laboratory, LSI Medience Corp., for conducting the functional observational battery experiment.

References

Autry AE, Adachi M, Nosyreva E et al (2011) NMDA receptor blockade at rest triggers rapid behavioural antidepressant responses. Nature 475:91–95

Becker A, Grecksch G (2004) Ketamine-induced changes in rat behaviour: a possible animal model of schizophrenia. Test of predictive validity. Prog Neuropsychopharmacol Biol Psychiatry 28:1267–1277

Becker A, Peters B, Schroeder H et al (2003) Ketamine-induced changes in rat behaviour: a possible animal model of schizophrenia. Prog Neuropsychopharmacol Biol Psychiatry 27:687–700

Berman RM, Cappiello A, Anand A et al (2000) Antidepressant effects of ketamine in depressed patients. Biol Psychiatry 47:351–354

Caffino L, Piva A, Mottarlini F et al (2018) Ketamine self-administration elevates αCaMKII autophosphorylation in mood and reward-related brain regions in rats. Mol Neurobiol 55:5453–5461

Castaneda AE, Tuulio-Henriksson A, Marttunen M et al (2008) A review on cognitive impairments in depressive and anxiety disorders with a focus on young adults. J Affect Disord 106:1–27

Cottone P, Iemolo A, Narayan AR et al (2013) The uncompetitive NMDA receptor antagonists ketamine and memantine preferentially increase the choice for a small, immediate reward in low-impulsive rats. Psychopharmacology (Berl) 226:127–138

De Luca MT, Badiani A (2011) Ketamine self-administration in the rat: evidence for a critical role of setting. Psychopharmacology (Berl) 214:549–556

Donegan JJ, Lodge DJ (2017) Hippocampal perineuronal nets are required for the sustained antidepressant effect of ketamine. Int J Neuropsychopharmacol 20:354–358

Duan TT, Tan JW, Yuan Q et al (2013) Acute ketamine induces hippocampal synaptic depression and spatial memory impairment through dopamine D1/D5 receptors. Psychopharmacology (Berl) 228:451–461

Featherstone RE, M Tatard-Leitman V, Suh JD et al (2013) Electrophysiological and behavioral responses to ketamine in mice with reduced Akt1 expression. Psychopharmacology (Berl) 227:639–649

Fitzgerald PJ, Yen JY, Watson BO (2019) Stress-sensitive antidepressant-like effects of ketamine in the mouse forced swim test. PLoS One 14:e0215554

Fukumoto K, Toki H, Iijima M et al (2017) Antidepressant potential of (R)-ketamine in rodent models: comparison with (S)-ketamine. J Pharmacol Exp Ther 361:9–16

Getachew B, Aubee JI, Schottenfeld RS et al (2018) Ketamine interactions with gut-microbiota in rats: relevance to its antidepressant and anti-inflammatory properties. BMC Microbiol 18:222

Grady SE, Marsh TA, Tenhouse A et al (2018) Ketamine for the treatment of major depressive disorder and bipolar depression: a review of the literature. Ment Health Clin 7:16–23

Hayase T, Yamamoto Y, Yamamoto K (2006) Behavioral effects of ketamine and toxic interactions with psychostimulants. BMC Neurosci 7:25

Hillhouse TM, Porter JH, Negus SS (2014) Dissociable effects of the noncompetitive NMDA receptor antagonists ketamine and MK-801 on intracranial self-stimulation in rats. Psychopharmacology (Berl) 231:2705–2716

Honsberger MJ, Taylor JR, Corlett PR (2015) Memories reactivated under ketamine are subsequently stronger: a potential pre-clinical behavioral model of psychosis. Schizophr Res 164:227–233

Joules R, Doyle OM, Schwarz AJ et al (2015) Ketamine induces a robust whole-brain connectivity pattern that can be differentially modulated by drugs of different mechanism and clinical profile. Psychopharmacology (Berl) 232:4205–4218

Ju LS, Yang JJ, Lei L et al (2017) The combination of long-term ketamine and extinction training contributes to fear erasure by BDNF methylation. Front Cell Neurosci 11:100

Juven-Wetzler A, Cohen H, Kaplan Z et al (2014) Immediate ketamine treatment does not prevent posttraumatic stress responses in an animal model for PTSD. Eur Neuropsychopharmacol 24:469–479

Koek W, Woods JH, Ornstein P (1987) A simple and rapid method for assessing similarities among directly observable behavioral effects of drugs: PCP-like effects of 2-amino-5-phosphonovalerate in rats. Psychopharmacology (Berl) 91:297–304

Kunugi H, Tanaka M, Hori H et al (2007) Prepulse inhibition of acoustic startle in Japanese patients with chronic schizophrenia. Neurosci Res 59:23–28

Lapiz MN, Morilak DA (2006) Noradrenergic modulation of cognitive function in rat medial prefrontal cortex as measured by attentional set shifting capability. Neuroscience 137:1039–1049

Leander JD (1989) Tricyclic antidepressants block N-methyl-D-aspartic acid-induced lethality in mice. Br J Pharmacol 96:256–258

Li N, Liu RJ, Dwyer JM et al (2011) Glutamate N-methyl-D-aspartate receptor antagonists rapidly reverse behavioral and synaptic deficits caused by chronic stress exposure. Biol Psychiatry 69:754–761

Lin JC, Chan MH, Lee MY et al (2016) N,N-dimethylglycine differentially modulates psychotomimetic and antidepressant-like effects of ketamine in mice. Prog Neuropsychopharmacol Biol Psychiatry 71:7–13

Meliska CJ, Greenberg AJ, Trevor AJ (1980) The effects of ketamine enantiomers on schedule-controlled behavior in the rat. J Pharmacol Exp Ther 212:198–202

Miller OH, Yang L, Wang CC et al (2014) GluN2B-containing NMDA receptors regulate depression-like behavior and are critical for the rapid antidepressant actions of ketamine. Elife 3:e03581

Moghaddam B, Adams B, Verma A et al (1997) Activation of glutamatergic neurotransmission by ketamine: a novel step in the pathway from NMDA receptor blockade to dopaminergic and cognitive disruptions associated with the prefrontal cortex. J Neurosci 17:2921–2927

Nishimura M, Sato K, Okada T et al (1998) Ketamine inhibits monoamine transporters expressed in human embryonic kidney 293 cells. Anesthesiology 88:768–774

Orser BA, Pennefather PS, MacDonald JF (1997) Multiple mechanisms of ketamine blockade of N-methyl-D-aspartate receptors. Anesthesiology 86:903–917

Patton MS, Lodge DJ, Morilak DA et al (2017) Ketamine corrects stress-induced cognitive dysfunction through JAK2/STAT3 signaling in the orbitofrontal cortex. Neuropsychopharmacology 42:1220–1230

Pietersen CY, Bosker FJ, Doorduin J et al (2007) An animal model of emotional blunting in schizophrenia. PLoS One 2:e1360

Razoux F, Garcia R, Léna I (2007) Ketamine, at a dose that disrupts motor behavior and latent inhibition, enhances prefrontal cortex synaptic efficacy and glutamate release in the nucleus accumbens. Neuropsychopharmacology 32:719–727

Reynolds IJ, Miller RJ (1988) Tricyclic antidepressants block N-methyl-D-aspartate receptors: similarities to the action of zinc. Br J Pharmacol 95:95–102

Sanacora G, Treccani G, Popoli M (2012) Towards a glutamate hypothesis of depression: an emerging frontier of neuropsychopharmacology for mood disorders. Neuropharmacology 62:63–77

Sarkar A, Kabbaj M (2016) Sex differences in effects of ketamine on behavior, spine density, and synaptic proteins in socially isolated rats. Biol Psychiatry 80:448–456

Stoet G, Snyder LH (2006) Effects of the NMDA antagonist ketamine on task-switching performance: evidence for specific impairments of executive control. Neuropsychopharmacology 31:1675–1681

Strong CE, Schoepfer KJ, Dossat AM et al (2017) Locomotor sensitization to intermittent ketamine administration is associated with nucleus accumbens plasticity in male and female rats. Neuropharmacology 121:195–203

Suzuki T, Aoki T, Kato H et al (1999) Effects of the 5-HT(3) receptor antagonist ondansetron on the ketamine- and dizocilpine-induced place preferences in mice. Eur J Pharmacol 385:99–102

Suzuki T, Kato H, Aoki T et al (2000) Effects of the non-competitive NMDA receptor antagonist ketamine on morphine-induced place preference in mice. Life Sci 67:383–389

Swerdlow NR, Braff DL, Hartston H et al (1996) Latent inhibition in schizophrenia. Schizophr Res 20:91–103

Uribe E, Landaeta J, Wix R et al (2013) Memantine reverses social withdrawal induced by ketamine in rats. Exp Neurobiol 22:18–22

van der Kam EL, de Vry J, Tzschentke TM (2007) Effect of 2-methyl-6-(phenylethynyl) pyridine on intravenous self-administration of ketamine and heroin in the rat. Behav Pharmacol 18:717–724

Venniro M, Mutti A, Chiamulera C (2015) Pharmacological and non-pharmacological factors that regulate the acquisition of ketamine self-administration in rats. Psychopharmacology (Berl) 232:4505–4514

Walker AK, Budac DP, Bisulco S et al (2013) NMDA receptor blockade by ketamine abrogates lipopolysaccharide-induced depressive-like behavior in C57BL/6J mice. Neuropsychopharmacology 38:1609–1616

Wallace DL, Han MH, Graham DL et al (2009) CREB regulation of nucleus accumbens excitability mediates social isolation-induced behavioral deficits. Nat Neurosci 12:200–209

Wang N, Yu HY, Shen XF et al (2015) The rapid antidepressant effect of ketamine in rats is associated with down-regulation of pro-inflammatory cytokines in the hippocampus. Ups J Med Sci 120:241–248

Wenger GR, Dews PB (1976) The effects of phencyclidine, ketamine, delta-amphetamine and pentobarbital on schedule-controlled behavior in the mouse. J Pharmacol Exp Ther 196:616–624

WHO Expert Committee on Drug Dependence (2003) Pre-review of psychoactive substances ketamine. WHO technical report series, vol 915, p 15

Winger G, Palmer RK, Woods JH (1989) Drug-reinforced responding: rapid determination of dose-response functions. Drug Alcohol Depend 24:135–142

Winger G, Hursh SR, Casey KL et al (2002) Relative reinforcing strength of three N-methyl-D-aspartate antagonists with different onsets of action. J Pharmacol Exp Ther 301:690–697

Wright KN, Strong CE, Addonizio MN et al (2017) Reinforcing properties of an intermittent, low dose of ketamine in rats: effects of sex and cycle. Psychopharmacology (Berl) 234:393–401

Yang C, Shirayama Y, Zhang JC et al (2015) R-ketamine: a rapid-onset and sustained antidepressant without psychotomimetic side effects. Transl Psychiatry 5:e632

Yang C, Qu Y, Fujita Y et al (2017) Possible role of the gut microbiota-brain axis in the antidepressant effects of (R)-ketamine in a social defeat stress model. Transl Psychiatry 7:1294

Zanos P, Gould TD (2018) Mechanisms of ketamine action as an antidepressant. Mol Psychiatry 23:801–811

Zhang K, Dong C, Fujita Y et al (2018) 5-Hydroxytryptamine-independent antidepressant actions of (R)-ketamine in a chronic social defeat stress model. Int J Neuropsychopharmacol 21:157–163

Zhu CB, Lindler KM, Owens AW et al (2010) Interleukin-1 receptor activation by systemic lipopolysaccharide induces behavioral despair linked to MAPK regulation of CNS serotonin transporters. Neuropsychopharmacology 35:2510–2020

Antidepressant Actions of Ketamine and Its Two Enantiomers

Lijia Chang, Yan Wei, and Kenji Hashimoto

Abstract The *N*-methyl-D-aspartate receptor antagonist ketamine has been widely used as an off-label medication to treat depression because it elicits rapid and robust antidepressant effects in treatment-resistant patients with depression. (*R*,*S*)-ketamine is a racemic mixture containing equal amounts of (*R*)-ketamine (or arketamine) and (*S*)-ketamine (or esketamine). On March 5, 2019, the United States Food and Drug Administration approved an (*S*)-ketamine nasal spray for treatment-resistant depression. In contrast, (*R*)-ketamine has been reported to have a greater potency and longer-lasting antidepressant effects than (*S*)-ketamine in rodent models of depression. However, the precise mechanisms underlying the robust antidepressant effects of ketamine enantiomers remain unknown. In this chapter, we discuss recent findings on the antidepressant actions of two enantiomers of ketamine.

Keywords Arketamine · Ketamine · Esketamine

1 Introduction

Ketamine is a noncompetitive *N*-methyl-D-aspartate receptor (NMDAR) antagonist that has drawn significant attention in recent decades from scientists all over the world because of its robust antidepressant effects (Cusin 2019; Duman 2018; Hashimoto 2016b, 2017, 2019). Ketamine was first synthesized at Parke-Davis (Detroit, MI, USA) by Calvin Lee Stevens in 1962, with a study of ketamine's

L. Chang · K. Hashimoto (✉)
Division of Clinical Neuroscience, Chiba University Center for Forensic Mental Health, Chiba, Japan
e-mail: hashimoto@faculty.chiba-u.jp

Y. Wei
Division of Clinical Neuroscience, Chiba University Center for Forensic Mental Health, Chiba, Japan

Key Laboratory of Medical Electrophysiology of Ministry of Education and Medical Electrophysiological Key Laboratory of Sichuan Province, Collaborative Innovation Center for Prevention and Treatment of Cadiovascular Disease, Institute of Cardiovascular Research, Southwest Medical University, Luzhou, Sichuan, China

K. Hashimoto et al. (eds.), *Ketamine*,
https://doi.org/10.1007/978-981-15-2902-3_7

dissociative anesthetics effects in the prison population quickly following in 1964 (Cohen et al. 2018; Domino 2010; Li and Vlisides 2016). Afterward, ketamine's anesthetic effects were subsequently confirmed by other investigators, and the drug began to be employed as an anesthetic in humans and animals in 1966 (Domino 2010; Stevenson 2005). In view of ketamine's unique pharmacological properties of pain relief and sedation (Gao et al. 2016), it was approved formally for use by the United States (US) Food and Drug Administration (FDA) in 1970 and was placed by the World Health Organization (WHO) in 1985 onto an essential medicines list as an intravenous anesthetic (Gowda et al. 2016; WHO 2011).

Berman et al. (2000) first conducted a placebo-controlled study of ketamine application in eight patients with major depressive disorder (MDD), who were given a dosage of 0.5 mg/kg over 40 min; their depressive symptoms significantly improved within 3 days, while feelings of perceptual disturbances or euphoria occurred in the same patients after treatment (Berman et al. 2000). More generally, in these individuals, the drug produced rapid-acting and sustained antidepressant effects although psychotomimetic effects were also reported after a single infusion of ketamine (Berman et al. 2000). Subsequent studies replicated the noted robust antidepressant effects of ketamine in treatment-resistant patients with MDD and bipolar disorder (BD) (Zarate et al. 2006; Newport et al. 2015; Kishimoto et al. 2016).

Despite ketamine's powerful antidepressant effects in humans, there are some potential drawbacks that cannot be neglected. When used as an anesthetic for anesthesia in surgical procedures, acute and chronic pain management, and critical care (Kurdi et al. 2014), side effects such as hallucinations, agitation, confusion, and psychotomimetic effects resembling schizophrenia can follow (Domino 2010; Gao et al. 2016; Krystal et al. 1994; Kurdi et al. 2014). For some patients, ketamine also brings about cognitive impairments (Molero et al. 2018; Szlachta et al. 2017), urinary tract inflammation (Sihra et al. 2018), liver enzyme abnormalities (Zhao et al. 2018), and other symptoms such as an increase in the heart rate or blood pressure (Sheth et al. 2018). Further, considering the effects reported during recreational usage or the euphoric "dissociated" state caused by higher doses, there exists a potential for abuse no matter what the drug's original intended purpose in a situation is (Heal et al. 2018; Ivan Ezquerra-Romano et al. 2018; Liao et al. 2017; Morgan

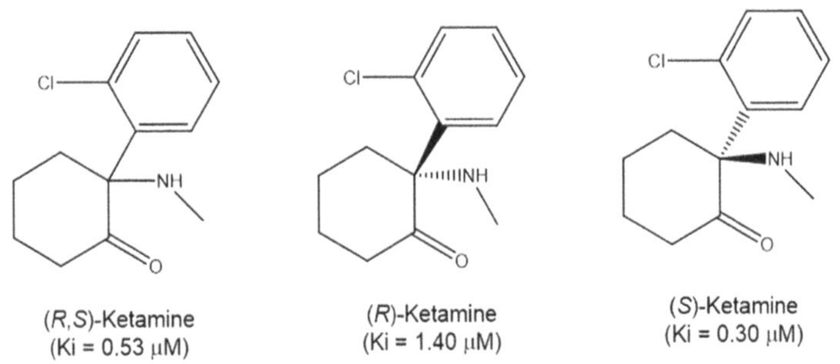

Fig. 1 Chemical structure of ketamine and its two enantiomers. The values in the parenthesis are the *K*i value for the NMDAR (Hashimoto 2019)

and Curran 2012; Tracy et al. 2017). At present, however, there are no superior antidepressants to ketamine that show robust antidepressant effects in patients with MDD (Duman 2018).

Ketamine is a racemic mixture containing equal amounts of (*R*)-ketamine (or arketamine) and (*S*)-ketamine (or esketamine) (Fig. 1) (Domino 2010). Different affinities and potencies of these two ketamine enantiomers exist. (*S*)-ketamine has about a three- to fourfold greater affinity for the NMDAR as compared with (*R*)-ketamine (Fig. 1). Conversely, (*S*)-ketamine holds approximately a three- to fourfold greater anesthetic potency and shows greater undesirable psychotomimetic side effects in comparison with (*R*)-ketamine (Domino 2010; Hashimoto 2016b, c). In prior research, (*R*)-ketamine showed a greater degree of potency and longer-lasting antidepressant effects than (*S*)-ketamine in several rodents model of depression (Fukumoto et al. 2017; Yang et al. 2015, 2018a; Zhang et al. 2014), and (*R*)-ketamine also did not present side effects on psychotomimetic actions or abuse liability in contrast with (*S*)-ketamine (Chang et al. 2019a; Yang et al. 2015). Third, (*R*)-ketamine is expected to constitute a safer choice than either (*R*,*S*)-ketamine or (*S*)-ketamine in consideration of its antidepressant actions (Chang et al. 2019a; Hashimoto 2016b, c). However, the precise molecular and cellular mechanisms underlying ketamine's antidepressant effects remain to be elucidated. To date, there is much debate ongoing about the molecular mechanisms underlying ketamine's antidepressant effects in humans and rodents alike. Therefore, it is necessary to summarize the current findings on ketamine's antidepressant effects to further the understanding of the neurobiology of ketamine's actions.

2 (*R*,*S*)-Ketamine's Antidepressant Effects

2.1 *Sumary of (*R,S*)-Ketamine's Antidepressant Effects*

Data from the WHO indicate that more than 300 million people of all ages are suffering from depression, which is the leading cause of disability worldwide, leading to a great expenditure of medical and health resources (WHO 2017). Currently available antidepressants have a delay period of approximately 3–4 weeks before they begin working, and about 30% of patients with depression are resistant entirely to these antidepressants (Zhang and Hashimoto 2019). Moreover, other NMDAR antagonists except ketamine have failed in clinical trials (Hashimoto 2019). Therefore, it is unlikely that NMDAR inhibition plays a key role in the antidepressant actions of ketamine.

(*R*,*S*)-Ketamine's antidepressant effects have been studied for almost 20 years after Berman et al.'s seminal report (Berman et al. 2000). In the following decade, most studies on patients with MDD have revealed promising prospects for the use of (*R*,*S*)-ketamine's antidepressant effects. Furthermore, clinical trials conducted on the subjects of ketamine's administration dosage, timing, route, curative effect, tolerability, safety, and antisuicidal effects have also been investigated. Even its use in different types of depression, including treatment-resistant depression, posttraumatic stress disorder (PTSD), and bipolar depression, has been explored in more recent studies (Bobo et al. 2016; Liriano et al. 2019; Schwartz et al. 2016; Zhang and Hashimoto 2019). Further, several rodent models of depression [i.e., the learned helplessness (LH) model, chronic unpredictable mild stress (CUMS) model, and chronic social defeat stress (CSDS) model] have been established for the purpose of testing (*R*,*S*)-ketamine's antidepressant effects (Krishnan and Nestler 2011). In short, an array of research modalities including original investigations, clinical trials, systematic reviews, and meta-analyses have been used to study (*R*,*S*)-ketamine's antidepressant effects (Krishnan and Nestler 2011; Sanacora et al. 2017; Short et al. 2018; Singh et al. 2017; Veraart et al. 2018).

2.2 (R,S)-Ketamine's Antidepressant Effects in Humans

In clinical trials, interventional studies have been adopted for evaluating the beneficial effects of (*R*,*S*)-ketamine. A number of randomized single-/double-blind placebo-/drug-controlled studies have been performed by many researchers to observe the antidepressant effects of (*R*,*S*)-ketamine in patients with MDD, who were often infused with only a single intravenous dose of 0.5 mg/kg. Berman et al. (2000) first reported the double-blinded and placebo-controlled study of (*R*,*S*)-ketamine. Then, Zarate et al. (2006) replicated that a high number (71%) of participants showed significant improvements in depressive symptoms within 24 h of drug administration, and 25% of the study subjects showed antidepressant effects for at least 1 week after a single dose of ketamine as part a randomized, placebo-controlled, double-blind crossover study.

A review of the ClinicalTrials.gov database found that about 160 applications from around the world for studying (*R*,*S*)-ketamine's antidepressant effect on patients with depression existed by the end of June 2019. Study subjects mainly included those with MDD with or without suicide ideation, treatment-resistant depression, BD, and PTSD. Studies chiefly involved a randomized single-blind or double-blind design and were placebo-controlled or midazolam-controlled interventions. Collectively, it seems that (*R*,*S*)-ketamine is viewed as a promising antidepressant for the treatment of severe depression. Systematic reviews and meta-analyses have also been adopted as investigational means by scientists for evaluating the impact of the antidepressant effect of (*R*,*S*)-ketamine. Such meta-analyses often include treatment-resistant patients with MDD or BD, and their data are mainly a collection of those from randomized controlled trials (RCTs) (Coyle and Laws 2015; Kishimoto et al. 2016; Newport et al. 2015; Serafini et al. 2014; Wilkinson et al. 2018).

(*R*,*S*)-Ketamine appears to exhibit rapid antidepressant and antisuicidal ideation effects in treatment-resistant patients with depression. Two midazolam-controlled randomized clinical trials conducted in depressed patients with low levels of suicidal ideation (Grunebaum et al. 2018; Murrough et al. 2013a) demonstrated rapid reductions in suicidal ideation and depressive symptoms within 24 h after treatment with (*R*,*S*)-ketamine as compared with in the midazolam treatment group (Grunebaum et al. 2018; Murrough et al. 2013a). (*R*,*S*)-Ketamine has similarly demonstrated early emerging evidence of having antisuicidal effects in depressed patients with suicidal ideation in other research that last from 1 day to 1 week (Bartoli et al. 2017; Reinstatler and Youssef 2015; Wilkinson et al. 2018).

In special studies on treatment-resistant depression, (*R*,*S*)-ketamine has been proposed as an effective antidepressant, but there is a lack of long-term data on sustained depression remission in patients (Serafini et al. 2014; Papadimitropoulou et al. 2017). When considering bipolar depression, (*R*,*S*)-ketamine was found to be an effective antidepressant (Parsaik et al. 2015), but there is also limited evidence to support the maintenance of a response for up to 24 h following a single intravenous dose of ketamine in bipolar depression, and (*R*,*S*)-ketamine did not show a statistical advantage in remission of bipolar depression (McCloud et al. 2015). Therefore, more RCT studies are needed to verify (*R*,*S*)-ketamine's antidepressant effect in different types of depression.

2.3 *Routes of Administration of (*R,S*)-Ketamine*

An understanding of the route of administration of (*R*,*S*)-ketamine is required to confirm its antidepressant effects. There exist a number of studies on routes of administration of (*R*,*S*)-ketamine, including intravenous, intramuscular, intranasal, subcutaneous, oral, sublingual, transmucosal, and intrarectal, and the bioavailability of these forms of administration varies widely as determined by the absorption of pharmacokinetics although the intravenous route is the most commonly used in patients with depression at this time (Andrade 2017; Hashimoto 2019; Zanos et al. 2018; Zhang and Hashimoto 2019). The bioavailability of intranasal administration of (*R*,*S*)-ketamine is lower than that of either intravenous or intramuscular administration (Hashimoto 2019; Zhang and Hashimoto 2019). In an effort to assess the safety and efficacy of (*R*,*S*)-ketamine, Andrade (2017) suggested that the infusion of ketamine by the subcutaneous, intranasal, and oral routes is worth further study based on their clinical practicability. Retrospective data from 22 patients given oral ketamine at a dose of 50 mg per 3 days, which was titrated up by 25 mg per 3 days, showed that 30% of patients with treatment-resistant depression responded to the oral (*R*,*S*)-ketamine and approximately 70% patients had no improvement in mood symptoms (Al Shirawi et al. 2017). Rosenblat et al. (2019) published a systematic review involving a small number of clinical studies that suggested that oral ketamine could produce significant antidepressant effects, but the drug's antisuicidal effects and efficacy remain undetermined in patients with treatment-resistant depression. Andrade (2019) suggested that oral ketamine should be evaluated by higher-quality studies in the future. Elsewhere, intranasal administration of a single dose of 50 mg of (*R*,*S*)-ketamine in a 2014 randomized, double-blind crossover study led to depressive symptoms being significantly improved in 44% of patients (8/18) after 24 h, with minimal psychosis and dissociation observed in the study (Lapidus et al. 2014). Thus, multiple routes other than the intravenous infusion of ketamine are under consideration in research.

2.4 (R,S)-Ketamine's Antidepressant Effects After Repeated Infusion

Repeated ketamine infusion is very important for maintenance therapy of patients with treatment-resistant depression (Murrough et al. 2013b; Strong and Kabbaj 2018). In a randomized controlled trial, the infusion of ketamine with three methods, including single, repeated, and maintenance, was performed to observe (*R*,*S*)-ketamine's antidepressant effects in treatment-resistant patients; results suggested that a single intravenous dose (0.5 mg/kg) over 40 min significantly improved depressive symptoms after 24 h of infusion. Elsewhere, six repeated infusions worked for a time in 59% of participants whose response criteria met the clinical needs, while these participants had no further benefit from weekly maintenance infusions based on the Montgomery–Åsberg Depression Rating Scale (MADRS) scores (Phillips et al. 2019). In addition, the efficiency of (*R*,*S*)-ketamine's antidepressant effects were also investigated in patients with PTSD treated by eight repeated ketamine infusions done over 4 weeks (Abdallah et al. 2019) and in those with unipolar and bipolar depressive disorder with current suicidal ideation treated by six repeated ketamine infusions performed during a 12-day period (Zhan et al. 2019; Zheng et al. 2018); these studies supported that the rapid and robust antidepressant effects of (*R*,*S*)-ketamine can be cumulative and sustained following repeated infusion. Despite ketamine's potential for abuse and psychotomimetic effects, it represents a new drug option for antidepressant treatment. Thus, further investigation of (*R*,*S*)-ketamine's antidepressant effects is necessary.

2.5 (R,S)-Ketamine's Antidepressant Effects in Rodents

Animal models of depression were adopted by researchers to investigate (*R*,*S*)-ketamine's antidepressant effects. Animal models of depression have previously been used for the development of new antidepressants (Fernando and Robbins 2011; Pittenger et al. 2007). In a shock-induced depression model, Chaturvedi et al. (1999) reported that the antidepressant effects of (*R*,*S*)-ketamine at a dose range of 2.5–10 mg/kg significantly increased ambulation and rearing in the open field test and attenuated the immobility time during the forced swimming test as compared with control mice who received shock. In a rat LH model, Koike et al. (2011) reported that (*R*,*S*)-ketamine (10 mg/kg) exerted rapid and sustained antidepressant effects for at least 72 h after treatment, in which the number of escape failures significantly decreased in the LH paradigm and the immobility time during the tail suspension test was significantly reduced in comparison with in the vehicle-treated group. In a CUMS model of either mice or rats, (*R*,*S*)-ketamine at a single dose of 10 mg/kg elicited rapid-onset and long-lasting antidepressant effects as evaluated by the behavioral tests of forced swimming, tail suspension, and sucrose preference; even the ameliorated anhedonia lasted for 8 days in this model (Ma et al. 2013; Sun et al. 2016). Hollis and Kabbaj (2014) suggested in their research that CSDS model is a valuable research tool for investigating possible causes and treatments for human depression. In a CSDS model, as compared with the TrkB agonists 7,8-dihydroxyflavone and TrkB antagonist ANA-12, (*R*,*S*)-ketamine showed longer-lasting antidepressant effects for 7 days (Zhang et al. 2015). In contrast, however, Donahue et al. (2014) reported that a single administration of ketamine (10 mg/kg) showed a rapid antidepressant effect in a CSDS model, attenuating social avoidance in the social interaction test but having no effect on anhedonia in the intracranial self-stimulation test (Donahue et al. 2014).

In addition, inflammation-induced mice or rat depression models, such as those using lipopolysaccharides or complete Freund's adjuvant, can also be useful in observing (*R*,*S*)-ketamine's antidepressant effects, in that depression-like behaviors were often determined by forced swimming test; sucrose preference test; and the pro-inflammatory cytokine levels of interleukin (IL)-1β, IL-6, and tumor necrosis factor-α (Ji et al. 2019; Remus and Dantzer 2016; Reus et al. 2017; Zhang et al. 2016). However, (*R*,*S*)-ketamine's potential psychotomimetic and other unwanted side effects have also been well-documented in these animal models, prompting behaviors such as hyperlocomotion, prepulse inhibition deficits, conditioned place preference, schizophrenia-like psychotic symptoms, and novel object recognition impairment (Becker et al. 2003; Chan et al. 2013; Chang et al. 2019a; Giorgetti et al. 2015; Imre et al. 2006; Lahti et al. 1995; Ma and Leung 2018; Yang et al. 2015; Zanos et al. 2017), which may pose a major barrier to further clinical study in patients with MDD.

3 (*S*)-Ketamine's Antidepressant Effects

3.1 Summary of (S)-Ketamine's Antidepressant Effects

(*S*)-Ketamine is an isomer of (*R*,*S*)-ketamine, which was introduced for medical use in analgesia and anesthesia in the early 1990s (Esketamine 2006; Trimmel et al. 2018). The US FDA originally stated that the stereoisomers of (*R*,*S*)-ketamine had not received enough attention for its commercial development in 1992 (Luft and Mendes 2005). Now, (*S*)-ketamine has been brought into greater focus over (*R*,*S*)-ketamine for its antidepressant effects in patients with treatment-resistant depression, due to its incorporation into a nasal spray approved by the US FDA on March

5, 2019 (Canady 2019; US Food and Drug Administration 2019). The evaluation of the (*S*)-ketamine nasal spray was conducted by way of four phase III RCTs, including three short-term (4-week) clinical trials and a withdrawal maintenance-of-effect trial, where patients experienced many side effects including disassociation, dizziness, nausea, sedation, vertigo, anxiety, and lethargy (US Food and Drug Administration 2019). Therefore, there are many challenges in the study of (*S*)-ketamine, which cannot meet the needs of all patients with MDD, and even the antidepressant efficiency and the side effects of (*S*)-ketamine are still under discussion. On the other hand, the existing literature or studies concerning the use of (*S*)-ketamine in patients with MDD number far less than those on the topic of (*R*,*S*)-ketamine, but the latter is still not yet approved by FDA for treatment via any route of administration.

In a case report, Paslakis et al. (2010) reported their results obtained with using off-label oral (*S*)-ketamine (1.25 mg/kg) within 14 days: two patients with treatment-resistant depression showed rapid and sustained improvement in their psychopathology within the first week, while another two patients did not respond to the drug throughout the entire treatment period. In addition, several studies have concluded that (*S*)-ketamine produces similar antidepressant effects to those of (*R*,*S*)-ketamine (Paul et al. 2009) but is better-tolerated than (*R*,*S*)-ketamine, yet severe psychotomimetic effects with (*S*)-ketamine have also been reported (Correia-Melo et al. 2017; Paul et al. 2009). It is worth mentioning that a case series of repeated (*S*)-ketamine (0.25 mg/kg over 40 min) off-label use demonstrated an improvement in depressive symptoms in 50% of the patients following infusion within 1 or 2 weeks, although 25% of the patients suffered from dissociative symptoms (Segmiller et al. 2013). Another case series reported that the safety and efficacy of (*S*)-ketamine were well demonstrated in patients with treatment-resistant depression (TRD); still, the potential for associated psychotic symptoms can leave some patients out of consideration (Ajub and Lacerda 2018). Thus, additional case studies are needed to review the efficacy and psychosis effects of (*S*)-ketamine.

According to clinical trials records from ClinicalTrials.gov, about 18 applications for studying (*S*)-ketamine's antidepressant effect in patients with depression have been uploaded as of April 1, 2019. Some clinical trials have been completed, while others are currently recruiting patients. The route of administration in these clinical trials is mainly intranasal injection, for evaluating the efficacy and safety in patients with MDD or TRD. Therapeutic regimens in these clinical trials also include single and repeated administration. Singh et al. (2016) found in a double-blind, multicenter, randomized, placebo-controlled trial involving patients with TRD that 67% and 64% of the patients, respectively, responded to (*S*)-ketamine given at a single dose of 0.2 mg/kg or 0.4 mg/kg over 40 min within 1 day. Another clinical trial of a head-to-head study compared the antidepressant actions of (*S*)-ketamine and (*R*,*S*)-ketamine in 96 individuals with TRD given a single dose of (*S*)-ketamine (0.25 mg/kg) or (*R*,*S*)-ketamine (0.5 mg/kg) over 40 min (Correia-Melo et al. 2018). The authors considered their study as the best way to evaluate the efficacy and safety of (*S*)-ketamine and (*R*,*S*)-ketamine, as adverse psychotomimetic effects between the two compounds may be different (Correia-Melo et al.

2018). Clinical trials still need to provide more powerful evidence to assist the clinical application of (*S*)-ketamine in depression or TRD.

Other research on (*S*)-ketamine's antidepressant effects mainly involves combination therapy. Bartova et al. (2015) investigated combined intravenous therapy of (*S*)-ketamine and oral tranylcypromine in two patients with multi-treatment-resistant depression, showing a confusing outcome of evident antisuicidal effects but no detailed improvement in antidepressant effectiveness. Electroconvulsive therapy (ECT) plus (*S*)-ketamine has also been applied in three patients with TRD. Kallmunzer et al. (2016) reported that all subjects showed remission in terms of suicidal ideation and no serious side effects during treatment. In contrast, another clinical trial suggested that (*S*)-ketamine (0.4 mg/kg in a bolus) as adjuvant therapy with propofol made no contribution to the enhanced role of ECT in patients with resistance to antidepressants, even going as far as to cause adverse effects like posttreatment disorientation and restlessness (Jarventausta et al. 2013). Although ECT can help to remedy depression, there is still a lack of clear evidence (Read et al. 2019). Therefore, there is no valid data to prove the combination of (*S*)-ketamine and ECT is beneficial, especially given that the antidepressant effects of (*R*,*S*)-ketamine are more potent than those of ECT (Muller et al. 2016).

In humans, the idea of using (*S*)-ketamine as an antidepressant is relatively new and its intranasal treatment modality has recently been receiving a lot of attention. However, many challenges need to be faced in future study. For instance, how about its optimal dose and route of administration? How to solve its side effects during treatment? How long does it sustain antidepressant effects? More importantly, how does it work as an antidepressant? These questions remain without an answer at this time.

3.2 (S)-Ketamine's Antidepressant Effects in Rodents

Animal experiments aimed at elucidating (*S*)-ketamine's antidepressant effect and its underlying mechanisms have also been conducted although no definitive answer was achieved. The current findings point out that (*S*)-ketamine works in a very complicated way, in that it can bind to multiple receptors in an organism, such as NMDAR, opioid receptors, monoamine receptors, adenosine receptors, α-amino-3-hydroxy-5-methyl-4-isoxazolepropionic acid (AMPAR) receptors, metabotropic glutamate receptors, L-type calcium channels, and other purinergic receptors (Kohrs and Durieux 1998; Trimmel et al. 2018). The prevailing view is that (*S*)-ketamine's antidepressant effect is related with the noncompetitive inhibition of NMDAR through binding to the phencyclidine (PCP) sites (Sinner and Graf 2008), due to its three- to fourfold higher affinity in comparison with (*R*)-ketamine (Domino 2010; Kohrs and Durieux 1998). Yet, (*S*)-ketamine dominates in the areas of anesthetic potency and undesirable psychotomimetic side effects over (*R*)-ketamine, in that seeing illusions, hearing and vision changes, and proprioception are related to

(*S*)-ketamine, while feelings of relaxation are connected with the actions of (*R*)-ketamine (Zanos et al. 2018).

Treccani et al. (2019) suggested that a single dose of (*S*)-ketamine (15 mg/kg) can induce acute antidepressant action in a rat model of depression, concluding first that dendritic spine density rapidly (<1 h) increases in brain structural changes and also that balancing the relationship between cofilin activity, homer3 levels, and the NMDAR subunits GluN2A and GluN2B plays a key role in the antidepressant effect. However, Ide et al. (2017), upon investigating the role of GluN2D in the antidepressant effects of ketamine enantiomers, suggested that such did not play a key role in the sustained antidepressant effects of (*R*,*S*)-ketamine and (*S*)-ketamine but rather in those of (*R*)-ketamine. Recent studies have indicated that (*S*)-ketamine elicits its acute and sustained antidepressant-like actions via a 5-hydroxytryptamine (5HT) receptor-dependent mechanism, which is confirmed via the forced swimming test by comparing the vortioxetine, fluoxetine, and (*S*)-ketamine in the serotonin system or by using the 5-HT_{1B} receptor agonist CP94253 in a genetic rat model (du Jardin et al. 2016, 2017).

Other studies of note regarding the mechanisms of (*S*)-ketamine's antidepressant actions are as follows. In a rat depression model of maternal deprivation, it was suggested that a single dose of (*S*)-ketamine showed long-term antidepressant effects through acting against neural damage induced by oxidative stress (Reus et al. 2015). In a long-term corticosterone infusion rat model of depression, it was suggested that (*S*)-ketamine decreased depressive-like behaviors for at least 4 weeks, but not in a manner related with rapid maturity of the new hippocampal neurons (Soumier et al. 2016). In a rat hippocampus study, Ardalan et al. (2017) put forward the notion that a single injection (15 mg/kg) of (*S*)-ketamine improved the immobility behavior during the modified forced swim test within 24 h after treatment, proving that astrocyte plasticity in the hippocampus has a bearing on the (*S*)-ketamine antidepressant effect in a rat genetic animal model of depression.

4 (*R*)-Ketamine's Antidepressant Effects

4.1 Summary of the Differences Between (R)-Ketamine and (S)-Ketamine

With the deepening and development of research, the focus on antidepressant effects will be quickly shifted toward (*R*)-ketamine, because its advantages are becoming more and more obvious. As compared with (*S*)-ketamine, (*R*)-ketamine has been proven to be more potent and have longer-lasting antidepressant effects and was deemed to be free of psychotomimetic side effects in many studies on rodent models of depression (Chang et al. 2019a; Fukumoto et al. 2017; Yang et al. 2015, 2016b; Zhang et al. 2014), signifying that (*R*)-ketamine will be one of the most promising antidepressants.

(*R*)-Ketamine is an isomer of (*R*,*S*)-ketamine. (*R*)-ketamine is a left-handed molecule, while (*S*)-ketamine is a right-handed molecule, per the carbon connections in either one or the other direction of nonsuperimposable mirror images, prompting a difference in the body metabolism (Calvey 1995; Kharasch and Labroo 1992). As compared with (*S*)-ketamine, (*R*)-ketamine has a lower affinity to the PCP binding site of the NMDAR, with a lower anesthetic potency (Thomson et al. 1985; White et al. 1985; Zeilhofer et al. 1992). In addition, (*R*)-ketamine could also weakly bind to the sigma receptor sites, without (*S*)-ketamine binding to these sites (Vollenweider et al. 1997). Research on the side effects of the two isomers showed that (*R*)-ketamine produces a feeling of relaxation or "well-being," while (*S*)-ketamine is mainly responsible for psychotomimetic effects, such as dissociations in hearing, vision, and proprioception (Hashimoto 2019; Vollenweider et al. 1997; Zanos et al. 2018). Moreover, Marietta et al. (1977) reported acute toxic effects of racemic ketamine and its two isomers at the dose of 40 mg/kg in male Sprague-Dawley rats based on median lethal dose values, indicating that (*R*)-ketamine is more secure than (*R*,*S*)-ketamine and (*S*)-ketamine, and its potential advantage is not a fatal drug according to the high dose and with less posthypnotic stimulation.

4.2 *Different Antidepressant Effects Between (*R*)-Ketamine and (*S*)-Ketamine*

Our group reported that (*R*)-ketamine has a greater potency and longer-lasting antidepressant roles than (*S*)-ketamine in rodent models of depression (Yang et al. 2015; Zhang et al. 2014). In addition, it was suggested that ketamine's side effects (e.g., psychotomimetic behaviors, neurotoxicity, abuse potential) may be associated with (*S*)-ketamine but not (*R*)-ketamine (Chang et al. 2019a; Hashimoto 2016a, b, c). Subsequently, different kinds of depression models with comparative objects and their relevant behavioral tests were employed to evaluate and identify the advantages of (*R*)-ketamine's antidepressant effects. Fukumoto et al. (2017) reported that (*R*)-ketamine can elicit longer-lasting antidepressant effects than (*S*)-ketamine in forced swimming and tail suspension tests, and (*R*)-ketamine showed a sustained antidepressant effect in a treatment-refractory model while (*S*)-ketamine did not. Moreover, Fukumoto et al. (2017) also suggested that (*R*)-ketamine's antidepressant effects may be related with AMPAR by testing the rats of a repeated corticosterone treatment model with an AMPAR antagonist, NBQX (Fukumoto et al. 2017). In a conscious positron-emission tomography study, a single infusion of (*S*)-ketamine but not (*R*)-ketamine caused dopamine release, indicating that (*S*)-ketamine-induced dopamine release may be associated with acute psychotomimetic and dissociative effects in humans (Hashimoto et al. 2017).

In addition, further research on the molecular mechanisms revealed more data about the differences between (*R*)-ketamine and (*S*)-ketamine. Specifically, the extracellular-signal-regulated kinase signaling pathway may play a role in

(*R*)-ketamine's antidepressant effects, while the mammalian target of rapamycin signaling pathway plays a role in the antidepressant effects of (*S*)-ketamine (Yang et al. 2018a). Nevertheless, glutamate is the primary excitatory neurotransmitter in the human brain in physiology (Jewett and Thapa 2019) and its participation in the role of ketamine's antidepressant effects should not be neglected in the following study, although we still cannot explain the real difference between (*R,S*)-ketamine and its two isomers (Zanos and Gould 2018). Even more important, the causes for the differences in antidepressant effects between (*R*)-ketamine and (*S*)-ketamine may be related with the gut microbiota–brain axis (Qu et al. 2017; Yang et al. 2017). As compared with (*S*)-ketamine, (*R*)-ketamine significantly attenuated the decrease in the levels of *Mollicutes* of susceptible mice in a CSDS model, and (*R*)-ketamine presented a stronger advantage than (*S*)-ketamine in reducing the levels of *Butyricimonas* (Yang et al. 2017).

4.3 (R)-*Ketamine's Antidepressant Effects*

(*R*)-ketamine has its own special features of antidepressant effects from the current study. As compared with the NMDAR partial agonist rapastinel, (*R*)-ketamine had longer-lasting antidepressant effects and significantly changed the brain-derived neurotrophic factor–TrkB signaling in a CSDS model (Yang et al. 2016a). Recent Ph3 data of rapastinel were negative (Hashimoto 2019).

In the light of the complex physiology and particularity that of NMDAR for regulating the electroneurographic signal in depression by voltage-gated ion channel influx, including calcium, sodium, and potassium (Cui et al. 2018; Yang et al. 2018b). Tian et al. (2018b) reported that the low-voltage-sensitive, T-type calcium channel blocker ethosuximide did not produce rapid or sustained antidepressant effects in a CSDS model, although (*R*)-ketamine showed rapid and sustained antidepressant effects in the same model (Tian et al. 2018b). Furthermore, Xiong et al. (2019) reported that the Kir4.1 inhibitors quinacrine and sertraline did not improve the depression-like behaviors during the tail suspension test and forced swimming test in a CSDS model, while (*R*)-ketamine produced rapid and sustained antidepressant effects in this model (Xiong et al. 2019).

Treatment with the 5-HT inhibitor para-chlorophenylalanine methyl ester hydrochloride also did not impact the antidepressant effects of (*R*)-ketamine in s CSDS model (Zhang et al. 2018). Furthermore, dopamine D_1 receptors may not play a major role in the antidepressant actions of (*R*)-ketamine, because pretreatment with the dopamine D_1 receptor antagonist SCH-23390 did not block the antidepressant effects of (*R*)-ketamine in the CSDS model (Chang et al. 2019b).

Given the role of gamma-aminobutyric acid receptors in depression (Zanos et al. 2017), both (*R*)-ketamine and MRK-016 (full inverse $GABA_A$ agonist) displayed rapid antidepressant effects, while only (*R*)-ketamine produced a longer-lasting antidepressant effect in a CSDS model (Xiong et al. 2018).

In addition, (*R*)-ketamine's antidepressant effects had regional differences in rat brains. A single bilateral infusion of (*R*)-ketamine was injected into the brain in a rat LH model of depression, showing that the infralimbic cortex of the medial prefrontal cortex (mPFC), dentate gyrus, and CA3 subregions of the hippocampus are related with (*R*)-ketamine's antidepressant effects as compared with other injection site areas including the mPFC subregion PrL, subregions of the shell and core in NAc, and BLA and CeA subregions of the amygdala, which had nothing to do with the antidepressant effects of (*R*)-ketamine (Shirayama and Hashimoto 2017). No toxicity was shown in the brain after repeated, intermittent administration with (*R*)-ketamine, with no loss in parvalbumin immunoreactivity in the brain region of mPFC and hippocampus observed, which was the opposite finding to those of (*S*)-ketamine (Yang et al. 2016b). Additionally, another interesting finding regarding neuropathological changes revealed that the neuronal injury marker heat shock protein HSP-70 was expressed in the rat retrosplenial cortex not because of (*R*)-ketamine but instead due to (*R*,*S*)-ketamine and (*S*)-ketamine (Tian et al. 2018a).

Based on the above findings, (*R*)-ketamine shows the rapid-acting and long-lasting antidepressant effects in rodents although its precise mechanism underlying these effects is still uncertain. A clinical trial of (*R*)-ketamine in humans is currently underway (Hashimoto 2019).

5 Conclusion

The discovery of the robust antidepressant actions of (*R*,*S*)-ketamine in depressed patients is serendipitous (Krystal et al. 2019). However, (*R*,*S*)-ketamine has not been approved by the US FDA for antidepressant treatment in clinical application because of a lack of patent. An (*S*)-ketamine nasal spray was approved by the FDA on March 5, 2019, although its side effects are known and it must be administrated under supervision in a certified doctor's office or outpatient clinic after treatment. It seems that antidepressant effects of (*S*)-ketamine in patients with MDD are less potent than those related with (*R*,*S*)-ketamine in clinical studies. Although there is no clinical trial of (*R*)-ketamine in patients with MDD, many preclinical data show its more potent antidepressant effects without side effects, suggesting that (*R*)-ketamine could be a safer antidepressant alternative to (*R*,*S*)-ketamine and (*S*)-ketamine. Due to its lower adverse side effects, we are looking forward to seeing more data on (*R*)-ketamine's antidepressant effects in patients with MDD and other psychiatric disorders.

Acknowledgments This study was supported by grant from AMED, Japan (to K.H., JP19dm0107119). Dr. Lijia Chang was supported by the Japan China Sasakawa Medical Fellowship (Tokyo, Japan). Dr. Yan Wei (Southwest Medical University, China) was supported by the China Scholarship Council (China).

Disclosure Statement: Dr. Hashimoto is the inventor of filed patent applications on "The use of *R*-ketamine in the treatment of psychiatric diseases" by the Chiba University. Dr. Hashimoto also declares that he has received research support and consultant from Dainippon Sumitomo, Otsuka, and Taisho.

References

Abdallah CG et al (2019) Repeated ketamine infusions for antidepressant-resistant PTSD: methods of a multicenter, randomized, placebo-controlled clinical trial. Contemp Clin Trials 81:11–18. https://doi.org/10.1016/j.cct.2019.04.009

Ajub E, Lacerda ALT (2018) Efficacy of esketamine in the treatment of depression with psychotic features: a case series. Biol Psychiatry 83:e15–e16. https://doi.org/10.1016/j.biopsych.2017.06.011

Al Shirawi MI, Kennedy SH, Ho KT, Byrne R, Downar J (2017) Oral ketamine in treatment-resistant depression: a clinical effectiveness case series. J Clin Psychopharmacol 37:464–467. https://doi.org/10.1097/JCP.0000000000000717

Andrade C (2017) Ketamine for depression, 4: in what dose, at what rate, by what route, for how long, and at what frequency? J Clin Psychiatry 78:e852–e857. https://doi.org/10.4088/JCP.17f11738

Andrade C (2019) Oral ketamine for depression, 2: practical considerations. J Clin Psychiatry 80. https://doi.org/10.4088/JCP.19f12838

Ardalan M, Rafati AH, Nyengaard JR, Wegener G (2017) Rapid antidepressant effect of ketamine correlates with astroglial plasticity in the hippocampus. Br J Pharmacol 174:483–492. https://doi.org/10.1111/bph.13714

Bartoli F, Riboldi I, Crocamo C, Di Brita C, Clerici M, Carra G (2017) Ketamine as a rapid-acting agent for suicidal ideation: a meta-analysis. Neurosci Biobehav Rev 77:232–236. https://doi.org/10.1016/j.neubiorev.2017.03.010

Bartova L, Vogl SE, Stamenkovic M, Praschak-Rieder N, Naderi-Heiden A, Kasper S, Willeit M (2015) Combination of intravenous S-ketamine and oral tranylcypromine in treatment-resistant depression: a report of two cases. Eur Neuropsychopharmacol 25:2183–2184. https://doi.org/10.1016/j.euroneuro.2015.07.021

Becker A, Peters B, Schroeder H, Mann T, Huether G, Grecksch G (2003) Ketamine-induced changes in rat behaviour: a possible animal model of schizophrenia. Prog Neuropsychopharmacol Biol Psychiatry 27:687–700. https://doi.org/10.1016/S0278-5846(03)00080-0

Berman RM, Cappiello A, Anand A, Oren DA, Heninger GR, Charney DS, Krystal JH (2000) Antidepressant effects of ketamine in depressed patients. Biol Psychiatry 47:351–354

Bobo WV, Vande Voort JL, Croarkin PE, Leung JG, Tye SJ, Frye MA (2016) Ketamine for treatment-resistant unipolar and bipolar major depression: critical review and implications for clinical practice. Depress Anxiety 33:698–710. https://doi.org/10.1002/da.22505

Calvey TN (1995) Isomerism and anaesthetic drugs. Acta Anaesthesiol Scand Suppl 106:83–90

Canady VA (2019) Nasal spray treatment for adults with TRD approved by FDA. Mental Health Wkly 29:4–5

Chan KW, Lee TM, Siu AM, Wong DP, Kam CM, Tsang SK, Chan CC (2013) Effects of chronic ketamine use on frontal and medial temporal cognition. Addict Behav 38:2128–2132. https://doi.org/10.1016/j.addbeh.2013.01.014

Chang L et al (2019a) Comparison of antidepressant and side effects in mice after intranasal administration of (R,S)-ketamine, (R)-ketamine, and (S)-ketamine. Pharmacol Biochem Behav 181:53–59. https://doi.org/10.1016/j.pbb.2019.04.008

Chang L et al (2019b) Lack of dopamine D1 receptors in the antidepressant actions of (R)-ketamine in a chronic social defeat stress model. Eur Arch Psychiatry Clin Neurosci. https://doi.org/10.1007/s00406-019-01012-1

Chaturvedi HK, Chandra D, Bapna JS (1999) Interaction between N-methyl-D-aspartate receptor antagonists and imipramine in shock-induced depression. Indian J Exp Biol 37:952–958

Cohen SP et al (2018) Consensus guidelines on the use of intravenous ketamine infusions for chronic pain from the American Society of Regional Anesthesia and Pain Medicine, the American Academy of Pain Medicine, and the American Society of Anesthesiologists. Reg Anesth Pain Med 43:521–546. https://doi.org/10.1097/AAP.0000000000000808

Correia-Melo FS, Silva SS, Araujo-de-Freitas L, Quarantini LC (2017) S-(+)-ketamine-induced dissociative symptoms as a traumatic experience in patients with treatment-resistant depression. Braz J Psychiatry 39:188–189. https://doi.org/10.1590/1516-4446-2016-2070

Correia-Melo FS et al (2018) Comparative study of esketamine and racemic ketamine in treatment-resistant depression: Protocol for a non-inferiority clinical trial. Medicine (Baltimore) 97:e12414. https://doi.org/10.1097/MD.0000000000012414

Coyle CM, Laws KR (2015) The use of ketamine as an antidepressant: a systematic review and meta-analysis. Hum Psychopharmacol 30:152–163. https://doi.org/10.1002/hup.2475

Cui Y et al (2018) Astroglial Kir4.1 in the lateral habenula drives neuronal bursts in depression. Nature 554:323–327. https://doi.org/10.1038/nature25752

Cusin C (2019) Ketamine as a rapid antidepressant. In: The Massachusetts General Hospital guide to depression. Springer, Berlin, pp 139–145

Domino EF (2010) Taming the ketamine tiger. 1965. Anesthesiology 113:678–684. https://doi.org/10.1097/ALN.0b013e3181ed09a2

Donahue RJ, Muschamp JW, Russo SJ, Nestler EJ, Carlezon WA Jr (2014) Effects of striatal DeltaFosB overexpression and ketamine on social defeat stress-induced anhedonia in mice. Biol Psychiatry 76:550–558. https://doi.org/10.1016/j.biopsych.2013.12.014

du Jardin KG, Liebenberg N, Muller HK, Elfving B, Sanchez C, Wegener G (2016) Differential interaction with the serotonin system by S-ketamine, vortioxetine, and fluoxetine in a genetic rat model of depression. Psychopharmacology (Berl) 233:2813–2825. https://doi.org/10.1007/s00213-016-4327-5

du Jardin KG, Liebenberg N, Cajina M, Muller HK, Elfving B, Sanchez C, Wegener G (2017) S-ketamine mediates its acute and sustained antidepressant-like activity through a 5-HT1B receptor dependent mechanism in a genetic rat model of depression. Front Pharmacol 8:978. https://doi.org/10.3389/fphar.2017.00978

Duman RS (2018) Ketamine and rapid-acting antidepressants: a new era in the battle against depression and suicide. F1000Res 7. https://doi.org/10.12688/f1000research.14344.1

Esketamine (2006) Drugs and lactation database (LactMed). National Library of Medicine (US), Bethesda, MD

Fernando AB, Robbins TW (2011) Animal models of neuropsychiatric disorders. Annu Rev Clin Psychol 7:39–61. https://doi.org/10.1146/annurev-clinpsy-032210-104454

Fukumoto K, Toki H, Iijima M, Hashihayata T, Yamaguchi JI, Hashimoto K, Chaki S (2017) Antidepressant potential of (R)-ketamine in rodent models: comparison with (S)-ketamine. J Pharmacol Exp Ther 361:9–16. https://doi.org/10.1124/jpet.116.239228

Gao M, Rejaei D, Liu H (2016) Ketamine use in current clinical practice. Acta Pharmacol Sin 37:865–872. https://doi.org/10.1038/aps.2016.5

Giorgetti R, Marcotulli D, Tagliabracci A, Schifano F (2015) Effects of ketamine on psychomotor, sensory and cognitive functions relevant for driving ability. Forensic Sci Int 252:127–142. https://doi.org/10.1016/j.forsciint.2015.04.024

Gowda MR, Srinivasa P, Kumbar PS, Ramalingaiah VH, Muthyalappa C, Durgoji S (2016) Rapid resolution of grief with IV infusion of ketamine: a unique phenomenological experience. Indian J Psychol Med 38:62–64. https://doi.org/10.4103/0253-7176.175121

Grunebaum MF et al (2018) Ketamine for rapid reduction of suicidal thoughts in major depression: a midazolam-controlled randomized clinical trial. Am J Psychiatry 175:327–335. https://doi.org/10.1176/appi.ajp.2017.17060647

Hashimoto K (2016a) Detrimental side effects of repeated ketamine infusions in the brain. Am J Psychiatry 173:1044–1045. https://doi.org/10.1176/appi.ajp.2016.16040411

Hashimoto K (2016b) Ketamine's antidepressant action: beyond NMDA receptor inhibition. Expert Opin Ther Targets 20:1389–1392. https://doi.org/10.1080/14728222.2016.1238899

Hashimoto K (2016c) Letter to the Editor: R-ketamine: a rapid-onset and sustained antidepressant without risk of brain toxicity. Psychol Med 46:2449–2451. https://doi.org/10.1017/S0033291716000969

Hashimoto K (2017) Rapid antidepressant activity of ketamine beyond NMDA receptor. In: The NMDA receptors. Springer, Berlin, pp 69–81

Hashimoto K (2019) Rapid-acting antidepressant ketamine, its metabolites and other candidates: A historical overview and future perspective. Psychiatry Clin Neurosci 73(10):613–627. https://doi.org/10.1111/pcn.12902

Hashimoto K, Kakiuchi T, Ohba H, Nishiyama S, Tsukada H (2017) Reduction of dopamine $D_{2/3}$ receptor binding in the striatum after a single administration of esketamine, but not *R*-ketamine: a PET study in conscious monkeys. Eur Arch Psychiatry Clin Neurosci 267:173–176

Heal DJ, Gosden J, Smith SL (2018) Evaluating the abuse potential of psychedelic drugs as part of the safety pharmacology assessment for medical use in humans. Neuropharmacology 142:89–115. https://doi.org/10.1016/j.neuropharm.2018.01.049

Hollis F, Kabbaj M (2014) Social defeat as an animal model for depression. ILAR J 55:221–232. https://doi.org/10.1093/ilar/ilu002

Ide S, Ikekubo Y, Mishina M, Hashimoto K, Ikeda K (2017) Role of NMDA receptor GluN2D subunit in the antidepressant effects of enantiomers of ketamine. J Pharmacol Sci 135:138–140. https://doi.org/10.1016/j.jphs.2017.11.001

Imre G, Salomons A, Jongsma M, Fokkema DS, Den Boer JA, Ter Horst GJ (2006) Effects of the mGluR2/3 agonist LY379268 on ketamine-evoked behaviours and neurochemical changes in the dentate gyrus of the rat. Pharmacol Biochem Behav 84:392–399. https://doi.org/10.1016/j.pbb.2006.05.021

Ivan Ezquerra-Romano I, Lawn W, Krupitsky E, Morgan CJA (2018) Ketamine for the treatment of addiction: evidence and potential mechanisms. Neuropharmacology 142:72–82. https://doi.org/10.1016/j.neuropharm.2018.01.017

Jarventausta K et al (2013) Effects of S-ketamine as an anesthetic adjuvant to propofol on treatment response to electroconvulsive therapy in treatment-resistant depression: a randomized pilot study. J ECT 29:158–161. https://doi.org/10.1097/YCT.0b013e318283b7e9

Jewett BE, Thapa B (2019) Physiology, NMDA Receptor. In: StatPearls. StatPearls, Treasure Island, FL

Ji M et al (2019) Acute ketamine administration attenuates lipopolysaccharide-induced depressive-like behavior by reversing abnormal regional homogeneity in the nucleus accumbens. Neuroreport 30:421–427. https://doi.org/10.1097/WNR.0000000000001219

Kallmunzer B, Volbers B, Karthaus A, Tektas OY, Kornhuber J, Muller HH (2016) Treatment escalation in patients not responding to pharmacotherapy, psychotherapy, and electro-convulsive therapy: experiences from a novel regimen using intravenous S-ketamine as add-on therapy in treatment-resistant depression. J Neural Transm (Vienna) 123:549–552. https://doi.org/10.1007/s00702-015-1500-7

Kharasch ED, Labroo R (1992) Metabolism of ketamine stereoisomers by human liver microsomes. Anesthesiology 77:1201–1207

Kishimoto T, Chawla JM, Hagi K, Zarate CA, Kane JM, Bauer M, Correll CU (2016) Single-dose infusion ketamine and non-ketamine N-methyl-d-aspartate receptor antagonists for unipolar and bipolar depression: a meta-analysis of efficacy, safety and time trajectories. Psychol Med 46:1459–1472. https://doi.org/10.1017/S0033291716000064

Kohrs R, Durieux ME (1998) Ketamine: teaching an old drug new tricks. Anesth Analg 87:1186–1193

Koike H, Iijima M, Chaki S (2011) Involvement of AMPA receptor in both the rapid and sustained antidepressant-like effects of ketamine in animal models of depression. Behav Brain Res 224:107–111. https://doi.org/10.1016/j.bbr.2011.05.035

Krishnan V, Nestler EJ (2011) Animal models of depression: molecular perspectives. Curr Top Behav Neurosci 7:121–147. https://doi.org/10.1007/7854_2010_108

Krystal JH et al (1994) Subanesthetic effects of the noncompetitive NMDA antagonist, ketamine, in humans. Psychotomimetic, perceptual, cognitive, and neuroendocrine responses. Arch Gen Psychiatry 51:199–214. https://doi.org/10.1001/archpsyc.1994.03950030035004

Krystal JH, Abdallah CG, Sanacora G, Charney DS, Duman RS (2019) Ketamine: a paradigm shift for depression research and treatment. Neuron 101:774–778. https://doi.org/10.1016/j.neuron.2019.02.005

Kurdi MS, Theerth KA, Deva RS (2014) Ketamine: current applications in anesthesia, pain, and critical care. Anesth Essays Res 8:283–290. https://doi.org/10.4103/0259-1162.143110

Lahti AC, Holcomb HH, Medoff DR, Tamminga CA (1995) Ketamine activates psychosis and alters limbic blood flow in schizophrenia. Neuroreport 6:869–872

Lapidus KA et al (2014) A randomized controlled trial of intranasal ketamine in major depressive disorder. Biol Psychiatry 76:970–976. https://doi.org/10.1016/j.biopsych.2014.03.026

Li L, Vlisides PE (2016) Ketamine: 50 years of modulating the mind. Front Hum Neurosci 10:612. https://doi.org/10.3389/fnhum.2016.00612

Liao Y, Tang YI, Hao W (2017) Ketamine and international regulations. Am J Drug Alcohol Abuse 43:495–504

Liriano F, Hatten C, Schwartz TL (2019) Ketamine as treatment for post-traumatic stress disorder: a review. Drugs Context 8:212305. https://doi.org/10.7573/dic.212305

Luft A, Mendes FF (2005) Low S(+) ketamine doses: a review. Rev Bras Anestesiol 55:460–469

Ma J, Leung LS (2018) Involvement of posterior cingulate cortex in ketamine-induced psychosis relevant behaviors in rats. Behav Brain Res 338:17–27. https://doi.org/10.1016/j.bbr.2017.09.051

Ma XC et al (2013) Long-lasting antidepressant action of ketamine, but not glycogen synthase kinase-3 inhibitor SB216763, in the chronic mild stress model of mice. PLoS One 8:e56053. https://doi.org/10.1371/journal.pone.0056053

Marietta MP, Way WL, Castagnoli N Jr, Trevor AJ (1977) On the pharmacology of the ketamine enantiomorphs in the rat. J Pharmacol Exp Ther 202:157–165

McCloud TL et al (2015) Ketamine and other glutamate receptor modulators for depression in bipolar disorder in adults. Cochrane Database Syst Rev:CD011611. https://doi.org/10.1002/14651858.CD011611.pub2

Molero P, Ramos-Quiroga J, Martin-Santos R, Calvo-Sánchez E, Gutiérrez-Rojas L, Meana JJ (2018) Antidepressant efficacy and tolerability of ketamine and esketamine: a critical review. CNS Drugs 32:411–420

Morgan CJA, Curran HV (2012) Ketamine use: a review. Addiction 107:27–38. https://doi.org/10.1111/j.1360-0443.2011.03576.x

Muller J, Pentyala S, Dilger J, Pentyala S (2016) Ketamine enantiomers in the rapid and sustained antidepressant effects. Ther Adv Psychopharmacol 6:185–192. https://doi.org/10.1177/2045125316631267

Murrough JW et al (2013a) Antidepressant efficacy of ketamine in treatment-resistant major depression: a two-site randomized controlled trial. Am J Psychiatry 170:1134–1142. https://doi.org/10.1176/appi.ajp.2013.13030392

Murrough JW et al (2013b) Rapid and longer-term antidepressant effects of repeated ketamine infusions in treatment-resistant major depression. Biol Psychiatry 74:250–256. https://doi.org/10.1016/j.biopsych.2012.06.022

Newport DJ et al (2015) Ketamine and other NMDA antagonists: early clinical trials and possible mechanisms in depression. Am J Psychiatry 172:950–966. https://doi.org/10.1176/appi.ajp.2015.15040465

Papadimitropoulou K, Vossen C, Karabis A, Donatti C, Kubitz N (2017) Comparative efficacy and tolerability of pharmacological and somatic interventions in adult patients with treatment-resistant depression: a systematic review and network meta-analysis. Curr Med Res Opin 33:701–711. https://doi.org/10.1080/03007995.2016.1277201

Parsaik AK, Singh B, Khosh-Chashm D, Mascarenhas SS (2015) Efficacy of ketamine in bipolar depression: systematic review and meta-analysis. J Psychiatr Pract 21:427–435. https://doi.org/10.1097/PRA.0000000000000106

Paslakis G, Gilles M, Meyer-Lindenberg A, Deuschle M (2010) Oral administration of the NMDA receptor antagonist S-ketamine as add-on therapy of depression: a case series. Pharmacopsychiatry 43:33–35. https://doi.org/10.1055/s-0029-1237375

Paul R, Schaaff N, Padberg F, Moller HJ, Frodl T (2009) Comparison of racemic ketamine and S-ketamine in treatment-resistant major depression: report of two cases. World J Biol Psychiatry 10:241–244. https://doi.org/10.1080/15622970701714370

Phillips JL et al (2019) Single, repeated, and maintenance ketamine infusions for treatment-resistant depression: a randomized controlled trial. Am J Psychiatry 176:401–409. https://doi.org/10.1176/appi.ajp.2018.18070834

Pittenger C, Sanacora G, Krystal JH (2007) The NMDA receptor as a therapeutic target in major depressive disorder. CNS Neurol Disord Drug Targets 6:101–115

Qu Y, Yang C, Ren Q, Ma M, Dong C, Hashimoto K (2017) Comparison of (R)-ketamine and lanicemine on depression-like phenotype and abnormal composition of gut microbiota in a social defeat stress model. Sci Rep 7:15725. https://doi.org/10.1038/s41598-017-16060-7

Read J, Cunliffe S, Jauhar S, McLoughlin DM (2019) Should we stop using electroconvulsive therapy? BMJ 364:k5233. https://doi.org/10.1136/bmj.k5233

Reinstatler L, Youssef NA (2015) Ketamine as a potential treatment for suicidal ideation: a systematic review of the literature. Drugs R D 15:37–43. https://doi.org/10.1007/s40268-015-0081-0

Remus JL, Dantzer R (2016) Inflammation models of depression in rodents: relevance to psychotropic drug discovery. Int J Neuropsychopharmacol 19:pyw028. https://doi.org/10.1093/ijnp/pyw028

Reus GZ et al (2015) A single dose of S-ketamine induces long-term antidepressant effects and decreases oxidative stress in adulthood rats following maternal deprivation. Dev Neurobiol 75:1268–1281. https://doi.org/10.1002/dneu.22283

Reus GZ et al (2017) Ketamine potentiates oxidative stress and influences behavior and inflammation in response to lipolysaccharide (LPS) exposure in early life. Neuroscience 353:17–25. https://doi.org/10.1016/j.neuroscience.2017.04.016

Rosenblat JD, Carvalho AF, Li M, Lee Y, Subramanieapillai M, McIntyre RS (2019) Oral ketamine for depression: a systematic review. J Clin Psychiatry 80. https://doi.org/10.4088/JCP.18r12475

Sanacora G et al (2017) A consensus statement on the use of ketamine in the treatment of mood disorders. JAMA Psychiat 74:399–405. https://doi.org/10.1001/jamapsychiatry.2017.0080

Schwartz J, Murrough JW, Iosifescu DV (2016) Ketamine for treatment-resistant depression: recent developments and clinical applications. Evid Based Ment Health 19:35–38. https://doi.org/10.1136/eb-2016-102355

Segmiller F et al (2013) Repeated S-ketamine infusions in therapy resistant depression: a case series. J Clin Pharmacol 53:996–998. https://doi.org/10.1002/jcph.122

Serafini G, Howland RH, Rovedi F, Girardi P, Amore M (2014) The role of ketamine in treatment-resistant depression: a systematic review. Curr Neuropharmacol 12:444–461. https://doi.org/10.2174/1570159x12666140619204251

Sheth MK, Brand A, Halterman J (2018) Ketamine-induced changes in blood pressure and heart rate in pre-hospital intubated patients. Adv J Grad Res 3:20–33

Shirayama Y, Hashimoto K (2017) Effects of a single bilateral infusion of R-ketamine in the rat brain regions of a learned helplessness model of depression. Eur Arch Psychiatry Clin Neurosci 267:177–182. https://doi.org/10.1007/s00406-016-0718-1

Short B, Fong J, Galvez V, Shelker W, Loo CK (2018) Side-effects associated with ketamine use in depression: a systematic review. Lancet Psychiatry 5:65–78. https://doi.org/10.1016/S2215-0366(17)30272-9

Sihra N, Ockrim J, Wood D (2018) The effects of recreational ketamine cystitis on urinary tract reconstruction—a surgical challenge. BJU Int 121:458–465. https://doi.org/10.1111/bju.14094

Singh JB et al (2016) Intravenous esketamine in adult treatment-resistant depression: a double-blind, double-randomization, placebo-controlled study. Biol Psychiatry 80:424–431. https://doi.org/10.1016/j.biopsych.2015.10.018

Singh I, Morgan C, Curran V, Nutt D, Schlag A, McShane R (2017) Ketamine treatment for depression: opportunities for clinical innovation and ethical foresight. Lancet Psychiatry 4:419–426. https://doi.org/10.1016/S2215-0366(17)30102-5

Sinner B, Graf BM (2008) Ketamine. Handb Exp Pharmacol:313–333. https://doi.org/10.1007/978-3-540-74806-9_15

Soumier A, Carter RM, Schoenfeld TJ, Cameron HA (2016) New hippocampal neurons mature rapidly in response to ketamine but are not required for its acute antidepressant effects on neophagia in rats. eNeuro 3. https://doi.org/10.1523/ENEURO.0116-15.2016

Stevenson C (2005) Ketamine: a review. Update Anaesth 20:25–29

Strong CE, Kabbaj M (2018) On the safety of repeated ketamine infusions for the treatment of depression: effects of sex and developmental periods. Neurobiol Stress 9:166–175. https://doi.org/10.1016/j.ynstr.2018.09.001

Sun HL et al (2016) Role of hippocampal p11 in the sustained antidepressant effect of ketamine in the chronic unpredictable mild stress model. Transl Psychiatry 6:e741. https://doi.org/10.1038/tp.2016.21

Szlachta M et al (2017) Effect of clozapine on ketamine-induced deficits in attentional set shift task in mice. Psychopharmacology (Berl) 234:2103–2112. https://doi.org/10.1007/s00213-017-4613-x

Thomson AM, West DC, Lodge D (1985) An N-methylaspartate receptor-mediated synapse in rat cerebral cortex: a site of action of ketamine? Nature 313:479–481

Tian Z, Dong C, Fujita A, Fujita Y, Hashimoto K (2018a) Expression of heat shock protein HSP-70 in the retrosplenial cortex of rat brain after administration of (R,S)-ketamine and (S)-ketamine, but not (R)-ketamine. Pharmacol Biochem Behav 172:17–21. https://doi.org/10.1016/j.pbb.2018.07.003

Tian Z, Dong C, Zhang K, Chang L, Hashimoto K (2018b) Lack of antidepressant effects of low-voltage-sensitive T-type calcium channel blocker ethosuximide in a chronic social defeat stress model: comparison with (R)-ketamine. Int J Neuropsychopharmacol 21:1031–1036. https://doi.org/10.1093/ijnp/pyy072

Tracy DK, Wood DM, Baumeister D (2017) Novel psychoactive substances: types, mechanisms of action, and effects. BMJ 356:i6848. https://doi.org/10.1136/bmj.i6848

Treccani G et al (2019) S-ketamine reverses hippocampal dendritic spine deficits in flinders sensitive line rats within 1 h of administration. Mol Neurobiol 56(11):7368–7379. https://doi.org/10.1007/s12035-019-1613-3

Trimmel H, Helbok R, Staudinger T, Jaksch W, Messerer B, Schochl H, Likar R (2018) S(+)-ketamine: Current trends in emergency and intensive care medicine. Wien Klin Wochenschr 130:356–366. https://doi.org/10.1007/s00508-017-1299-3

US Food and Drug Administration (2019) FDA approves new nasal spray medication for treatment-resistant depression; available only at a certified doctor's office or clinic. PressAnnouncements/ucm632761.htm

Veraart JKE, Smith-Apeldoorn SY, Trueman H, de Boer MK, Schoevers RA, McShane R (2018) Characteristics of patients expressing an interest in ketamine treatment: results of an online survey. BJPsych Open 4:389–392. https://doi.org/10.1192/bjo.2018.51

Vollenweider FX, Leenders KL, Oye I, Hell D, Angst J (1997) Differential psychopathology and patterns of cerebral glucose utilisation produced by (*S*)- and (*R*)-ketamine in healthy volunteers using positron emission tomography (PET). Eur Neuropsychopharmacol 7:25–38

White PF, Schuttler J, Shafer A, Stanski DR, Horai Y, Trevor AJ (1985) Comparative pharmacology of the ketamine isomers. Studies in volunteers. Br J Anaesth 57:197–203. https://doi.org/10.1093/bja/57.2.197

Wilkinson ST et al (2018) The effect of a single dose of intravenous ketamine on suicidal ideation: a systematic review and individual participant data meta-analysis. Am J Psychiatry 175:150–158. https://doi.org/10.1176/appi.ajp.2017.17040472

World Health Organization (2011) WHO model list of essential medicines: 17th list, March 2011

World Health Organization (2017) Depression and other common mental disorders: global health estimates (No. WHO/MSD/MER/2017.2). World Health Organization

Xiong Z, Zhang K, Ishima T, Ren Q, Chang L, Chen J, Hashimoto K (2018) Comparison of rapid and long-lasting antidepressant effects of negative modulators of alpha5-containing $GABA_A$ receptors and (*R*)-ketamine in a chronic social defeat stress model. Pharmacol Biochem Behav 175:139–145. https://doi.org/10.1016/j.pbb.2018.10.005

Xiong Z et al (2019) Lack of rapid antidepressant effects of Kir4.1 channel inhibitors in a chronic social defeat stress model: Comparison with (*R*)-ketamine. Pharmacol Biochem Behav 176:57–62. https://doi.org/10.1016/j.pbb.2018.11.010

Yang C et al (2015) R-ketamine: a rapid-onset and sustained antidepressant without psychotomimetic side effects. Transl Psychiatry 5:e632. https://doi.org/10.1038/tp.2015.136

Yang B et al (2016a) Comparison of R-ketamine and rapastinel antidepressant effects in the social defeat stress model of depression. Psychopharmacology (Berl) 233:3647–3657. https://doi.org/10.1007/s00213-016-4399-2

Yang C, Han M, Zhang JC, Ren Q, Hashimoto K (2016b) Loss of parvalbumin-immunoreactivity in mouse brain regions after repeated intermittent administration of esketamine, but not R-ketamine. Psychiatry Res 239:281–283. https://doi.org/10.1016/j.psychres.2016.03.034

Yang C, Qu Y, Fujita Y, Ren Q, Ma M, Dong C, Hashimoto K (2017) Possible role of the gut microbiota-brain axis in the antidepressant effects of (R)-ketamine in a social defeat stress model. Transl Psychiatry 7:1294. https://doi.org/10.1038/s41398-017-0031-4

Yang C, Ren Q, Qu Y, Zhang JC, Ma M, Dong C, Hashimoto K (2018a) Mechanistic target of rapamycin-independent antidepressant effects of (R)-ketamine in a social defeat stress model. Biol Psychiatry 83:18–28. https://doi.org/10.1016/j.biopsych.2017.05.016

Yang Y, Cui Y, Sang K, Dong Y, Ni Z, Ma S, Hu H (2018b) Ketamine blocks bursting in the lateral habenula to rapidly relieve depression. Nature 554:317–322. https://doi.org/10.1038/nature25509

Zanos P, Gould TD (2018) Intracellular signaling pathways involved in (s)- and (r)-ketamine antidepressant actions. Biol Psychiatry 83:2–4. https://doi.org/10.1016/j.biopsych.2017.10.026

Zanos P, Nelson ME, Highland JN, Krimmel SR, Georgiou P, Gould TD, Thompson SM (2017) A negative allosteric modulator for alpha5 subunit-containing GABA receptors exerts a rapid and persistent antidepressant-like action without the side effects of the NMDA receptor antagonist ketamine in mice. eNeuro 4. https://doi.org/10.1523/ENEURO.0285-16.2017

Zanos P et al (2018) Ketamine and ketamine metabolite pharmacology: insights into therapeutic mechanisms. Pharmacol Rev 70:621–660. https://doi.org/10.1124/pr.117.015198

Zarate CA Jr et al (2006) A randomized trial of an N-methyl-D-aspartate antagonist in treatment-resistant major depression. Arch Gen Psychiatry 63:856–864. https://doi.org/10.1001/archpsyc.63.8.856

Zeilhofer HU, Swandulla D, Geisslinger G, Brune K (1992) Differential effects of ketamine enantiomers on NMDA receptor currents in cultured neurons. Eur J Pharmacol 213:155–158

Zhan Y et al (2019) A preliminary study of anti-suicidal efficacy of repeated ketamine infusions in depression with suicidal ideation. J Affect Disord 251:205–212. https://doi.org/10.1016/j.jad.2019.03.071

Zhang K, Hashimoto K (2019) An update on ketamine and its two enantiomers as rapid-acting antidepressants. Expert Rev Neurother 19:83–92. https://doi.org/10.1080/14737175.2019.1554434

Zhang JC, Li SX, Hashimoto K (2014) R (-)-ketamine shows greater potency and longer lasting antidepressant effects than S (+)-ketamine. Pharmacol Biochem Behav 116:137–141. https://doi.org/10.1016/j.pbb.2013.11.033

Zhang JC et al (2015) Comparison of ketamine, 7,8-dihydroxyflavone, and ANA-12 antidepressant effects in the social defeat stress model of depression. Psychopharmacology (Berl) 232:4325–4335. https://doi.org/10.1007/s00213-015-4062-3

Zhang GF et al (2016) Acute single dose of ketamine relieves mechanical allodynia and consequent depression-like behaviors in a rat model. Neurosci Lett 631:7–12. https://doi.org/10.1016/j.neulet.2016.08.006

Zhang K, Dong C, Fujita Y, Fujita A, Hashimoto K (2018) 5-hydroxytryptamine-independent antidepressant actions of (R)-ketamine in a chronic social defeat stress model. Int J Neuropsychopharmacol 21:157–163. https://doi.org/10.1093/ijnp/pyx100

Zhao J, Wang Y, Wang D (2018) The effect of ketamine infusion in the treatment of complex regional pain syndrome: a systemic review and meta-analysis. Curr Pain Headache Rep 22:12. https://doi.org/10.1007/s11916-018-0664-x

Zheng W et al (2018) Rapid and longer-term antidepressant effects of repeated-dose intravenous ketamine for patients with unipolar and bipolar depression. J Psychiatr Res 106:61–68. https://doi.org/10.1016/j.jpsychires.2018.09.013

The Role of Gut Microbiota in the Antidepressant Effects of Ketamine

Yue Wang, Xiaolin Xu, Ailin Luo, and Chun Yang

Abstract In recent years, the prevalence, mental disability, and suicide rates of depression have been increasing without a corresponding significant change in cure rate, making depression the second largest disease burden worldwide. There is an urgent need to find more effective drugs and other therapeutic strategies. Accumulating evidence has revealed that ketamine elicits a fast-acting and sustained antidepressant effect, but the potential mechanisms underlying its antidepressant effects are not yet fully clear. Previous studies have indicated that ketamine's mechanism of action involves the inhibition of presynaptic and postsynaptic *N*-methyl-D-aspartate receptors (NMDARs) in GABAergic interneurons and the activation of postsynaptic α-amino-3-hydroxy-5-methyl-4-isoxazolepropionic acid receptors (AMPARs) and the brain-derived neurotrophic factor-tyrosine kinase receptor B (BDNF-TrkB) signaling pathway. Additionally, there is growing evidence that the gut microbiota may play a crucial role in the antidepressant effects of ketamine. In this chapter, we will discuss recent findings regarding the correlation between gut microbiota and the antidepressant effects of ketamine and their potential mechanisms of action. Further understanding of these pathways will likely lead to the development of novel and more effective treatments for depression.

Keywords Gut microbiota · Ketamine · Depression · Gut–brain axis · Probiotics

1 Depression Status

Depression is a common mood disorder illness characterized by low mood and loss of interest, usually also combined with anxiety (Malhi and Mann 2018). According to the World Health Organization in 2007, more than 300 million people suffer from

Y. Wang · X. Xu · A. Luo
Department of Anesthesiology, Tongji Medical College, Huazhong University of Science and Technology, Tongji Hospital, Wuhan, China

C. Yang (✉)
Department of Anesthesiology and Perioperative Medicine, The First Affiliated Hospital of Nanjing Medical University, Nanjing, China

K. Hashimoto et al. (eds.), *Ketamine*,
https://doi.org/10.1007/978-981-15-2902-3_8

depression worldwide and, as of 2017, new data showed that approximately 800,000 of those will commit suicide yearly. Depression incidence is rising (World Health Organization 2017), and since 2014, it has been found to account for the largest disease burden worldwide as measured by years lost to disability (Smith 2014). The incidence of depression in Asian countries such as China is about 3.02%, relatively low when compared to 22.5% in Afghanistan, 6.16% in Switzerland, and 4.45% in the United States (Smith 2014). This observation may be explained by the differential methods of diagnosis used in different countries masking real incidence rates, for example, by missed diagnosis due to symptoms being described as stomachaches or headaches (Smith 2014). Despite long-standing awareness of this disease, depression has recently gained more attention from society due to its high morbidity, rate of recurrence, disability, suicide rate, and low cure rate (Warden et al. 2007).

2 Depression Treatment

A number of drugs have been studied so far for the treatment of both fast-acting and long-lasting depression, especially in refractory disease (e.g., major depressive disorder (MDD) and bipolar disorder). Monoamine oxidase inhibitors (MAOIs), tricyclic antidepressants (TCAs), heterocyclic antidepressants (HCAs), and others were typically used for the treatment of depression. As new drug development evolved, antidepressants such as selective serotonin reuptake inhibitors (SSRIs) and selective norepinephrine reuptake inhibitors (SNRIs) became a more popular choice for the treatment of depression. It is well recognized that these drugs are effective to treat MDD, acting typically by increasing the levels of serotonin and norepinephrine. There is, however, a clear time lag of weeks to months from initial drug uptake to achieve therapeutic effects. Thus, these drugs are not effective in the treatment of the suicidal patients (Trivedi et al. 2006). Additionally, approximately 33% of depressed patients do not respond adequately to the currently available antidepressant drugs (Trivedi et al. 2006). Therefore, there is a pressing need for further investigation into this disease and for the development of more effective antidepressant drugs.

Preclinical and clinical studies have found that TCAs and SSRIs have neurochemical and behavioral similarities to *N*-methyl-D-aspartate receptor (NMDAR) antagonists, suggesting that the decrease in function of the NMDAR may be the major contributor to their antidepressant effect (Skolnick et al. 1996; Peyrovian et al. 2019). In 2000, Berman et al. conducted the first clinical study on the efficacy of subanesthetic ketamine in patients with major depression in a double-blinded study when compared to placebo controls and demonstrated that ketamine significantly reduced depression symptoms within 72 h of administration (Berman et al. 2000). Thus, ketamine, a neurotoxic and addictive non-barbiturate intravenous anesthetic, attracted the attention of the medical community. The rapid-acting and sustained antidepressant effect of ketamine were later confirmed in a larger, double-blinded, placebo-controlled study (Zarate et al. 2006). Since then, many other studies have reached similar conclusions on the effective antidepressant outturn of ketamine (Zarate et al. 2012; McGirr et al. 2015; Xu et al. 2016; Lee et al. 2015a;

Goitsuka et al. 1987; Murrough et al. 2013). More importantly, data also suggest that ketamine has a potent anti-suicide effect (Price et al. 2009; Murrough et al. 2015; Bartoli et al. 2017; Grunebaum et al. 2018).

The exact molecular and cellular mechanisms behind ketamine treatment are still not well understood. Ketamine is a racemic compound consisting of equal amounts of (*R*)-enantiomer and (*S*)-enantiomer (Hashimoto 2019). It is well known that the inhibition of NMDAR is a crucial step for its potent antidepressant effects, and it was demonstrated that (*S*)-ketamine has approximately three- to four-fold greater affinity for the NMDAR than (*R*)-ketamine. Hence, (*S*)-ketamine is generally considered as a better acting antidepressant. Despite (*S*)-ketamine nasal spray being approved by the Food and Drug Administration (FDA) in the USA for the treatment of depression, its conspicuous side effects due to the high affinity to NMDAR cannot be ignored, and this drug should be solely available from a certified doctor's office or clinic (Hashimoto 2019; Reardon 2019; U.S. Food and Drug Administration 2019). At present, there is no report of a clinical trial for (*R*)-ketamine in humans but several preclinical studies have shown that (*R*)-ketamine has a more potent antidepressant effect and lower psychotomimetic side effects than (*S*)-ketamine (Zhang et al. 2014; Yang et al. 2015; Fukumoto et al. 2017). These studies also demonstrated that (*R*)-ketamine is safer, suggesting that (*R*)-ketamine may be a more effective antidepressant (Yang et al. 2015). In addition, other emerging studies are finding that ketamine's metabolites (e.g., (*S*)-norketamine and (*2R,6R*)-hydroxynorketamine) may also show antidepressant effects (Yang et al. 2018a; Chou et al. 2018).

3 The Mechanisms of Action of Ketamine

Although the exact mechanisms of action ketamine and its metabolites are yet to be determined, current knowledge will be described below. (*S*)-ketamine and its metabolite (*S*)-norketamine mainly activate NMDAR and subsequently upregulate the brain-derived neurotrophic factor (BDNF)-TrkB-mTORC1 signaling pathway (Yang et al. 2018b). However, the antidepressant mechanism of (*R*)-ketamine and its metabolite (*2R,6R*)-hydroxynorketamine (HNK) is believed to induce a different pathway. (*R*)-ketamine can also activate NMDAR, but following its activation, it subsequently activates either the BNDF-TrkB or ERK signaling pathways, thereby promoting the synthesis of postsynaptic proteins (Yang et al. 2018b). Interestingly, (*2R,6R*)-HNK exerts its antidepressant effects by activating AMPAR and increasing the release of BNDF (Chou et al. 2018). Additionally, vascular endothelial growth factor (VEGF), hyperpolarization-activated cyclic nucleotide–gated channel 1 (HCN1), protein p11, and microRNAs can also play a role in the antidepressant effects of ketamine (Clark-Raymond and Halaris 2013; Deyama et al. 2019; Zhang et al. 2016; Sun et al. 2016; Bortolozzi et al. 2014).

As well as activating the glutamatergic system, ketamine's mechanisms of action could potentially also include the monoaminergic (e.g., 5-HT, dopamine and norepinephrine) and opioid systems (Williams et al. 2018; Zhang and Hashimoto 2019).

Ketamine does increase the release of 5-hydroxytryptamine (5-HT), dopamine (DA), and norepinephrine (NE), thereby promoting its antidepressant effects (Ago et al. 2019). Moreover, Williams et al. demonstrated that ketamine's acute antidepressant effect requires opioid system activation (Fig. 1) (Williams et al. 2018).

4 Changes in Gut Microbiota Can Influence the Antidepressant Effects of Ketamine

Gut microbiota is the name given to the trillions of microorganisms that coexist in the digestive tract of humans. It is formed by around 1000 different bacterial species and carries approximately 150 times more microbial genes than the human genome (Yang and Yu 2018), thereby being described as "the second genome" (Qin et al. 2010). There is growing evidence that gut microbiota plays a key role in depression (Dinan and Cryan 2019; Pennisi 2019; Stower 2019; Zhang et al. 2017). Hoban et al. reported significant depression-like behavior in rats following treatment with chronic antibiotics to deplete the gut microbiota (Hoban et al. 2016). Interestingly, ketamine was able to reverse some of the symptoms and further showed preventive effects, accompanied by changes in the composition of the gut microbiota (Yang et al. 2017b; Qu et al. 2017; Getachew et al. 2018; Mastrodonato et al. 2018).

We reported that phylum *Actinobacteria*, class *Coriobacteriia*, order *Clostridiales*, family *Prevotellaceae*, and genus *Alloprevotella* are independently correlated with forced swimming test (FST) immobility time, suggesting that these microorganisms could be responsible for the antidepressant effects seen in rodents. Subsequent to this study, phylum *Actinobacteria* and class *Coriobacteriia* were shown to be potential biomarkers for the antidepressant effect of ketamine in an inflammation model that used receiver operating characteristic (ROC) curve methodology for their analysis (Huang et al. 2019). We further demonstrated that gut microbiota may play a critical role in the antidepressant effects of (*R*)-ketamine in a mouse model of chronic social defeat stress (CSDS). Specifically, at the phylum level, CSDS-susceptible mice showed decreased levels of *Tenericutes* and elevated levels of *Actinobacteria*. At the class level, both ketamine enantiomers significantly dampened the numbers of *Deltaproteobacteria* that were originally increased in the CSDS-susceptible mice. Interestingly, (*R*)-ketamine but not (*S*)-ketamine significantly reversed the reduced numbers of *Mollicutes* in these mice. At the genus level, both ketamine enantiomers significantly dampened the reduced levels of *Butyricimonas* in susceptible mice, with (*R*)-ketamine showing higher potency than (*S*)-ketamine. We also found that (*R*)-ketamine significantly dampened the altered levels of *Bacteroidales*, *Clostridiales*, and *Ruminococcaceae* in CSDS-susceptible mice. At the genus level, it significantly dampened the increase in the level of *Clostridium* (Qu et al. 2017). Getachew et al. replicated these findings in rats, indicating that, at the genus level, ketamine noticeably amplified *Lactobacillus*, *Turicibacter*, and *Sarcina* by 3.3-, 26-, and 42-fold, respectively. By contrast, opportunistic pathogens *Mucispirillum* and *Ruminococcus* were reduced approximately

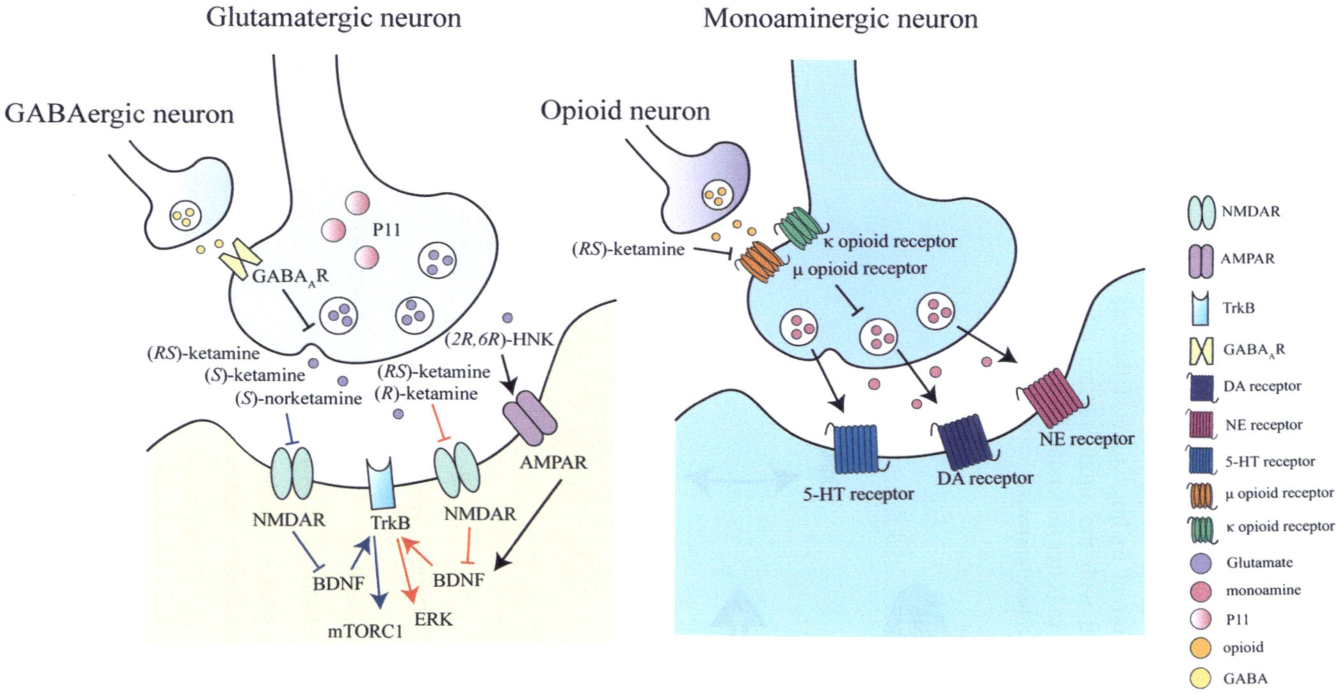

Fig. 1 The mechanisms involved in the antidepressant effects of ketamine. Ketamine is a racemic compound consisting of equal amounts of (*R*)-enantiomer and (*S*)-enantiomer, and its metabolites mainly include (*S*)-norketamine, (2*R*,6*R*)-HNK, etc. On the one hand, glutamatergic neurons are at the core of the antidepressant effects. (*S*)-ketamine and (*S*)-norketamine activate NMDAR, subsequently upregulate brain-derived neurotrophic factor (BDNF)-tyrosine kinase receptor B (TrkB)-mammalian target of rapamycin complex 1 (mTORC1) signaling. Moreover, (*R*)-ketamine mainly activates *N*-methyl-D-aspartic acid receptor (NMDAR), subsequently activates BNDF-TrkB signaling and extracellular signal-regulated kinase (ERK) in turns, thereby improving the depressive symptoms. However, (2*R*,6*R*)-ketamine activates AMPAR, subsequently promoting the release of BDNF. γ-Aminobutyric acid (GABAergic) interneurons and P11 also play a part in the effects. On the other hand, monoaminergic systems are of great importance. Ketamine exerts its antidepressant effects by inhibiting opioid receptors and promoting the release of 5-HT, dopamine (DA) and norepinephrine (NE)

2.6- and 26-fold, respectively, in the ketamine-treated group (Getachew et al. 2018). Collectively, these data show that the antidepressant effects of ketamine may be partially mediated by the composition of the gut microbiota (Yang et al. 2017b).

Despite these findings, increased fecal bacterial α-diversity was found in patients with MDD according to the Shannon index. The levels of *Bacteroidetes*, *Proteobacteria*, and *Actinobacteria* were significantly increased, whereas the levels of *Firmicutes* were significantly reduced in patients with MDD when compared with healthy individuals (Fig. 2) (Zheng et al. 2016; Jiang et al. 2015).

Taken together, the current findings suggest that the composition of the gut microbiome is associated with the antidepressant effects of ketamine. Further studies will be necessary to better understand what species of microbiota could be playing a key role in these mechanisms.

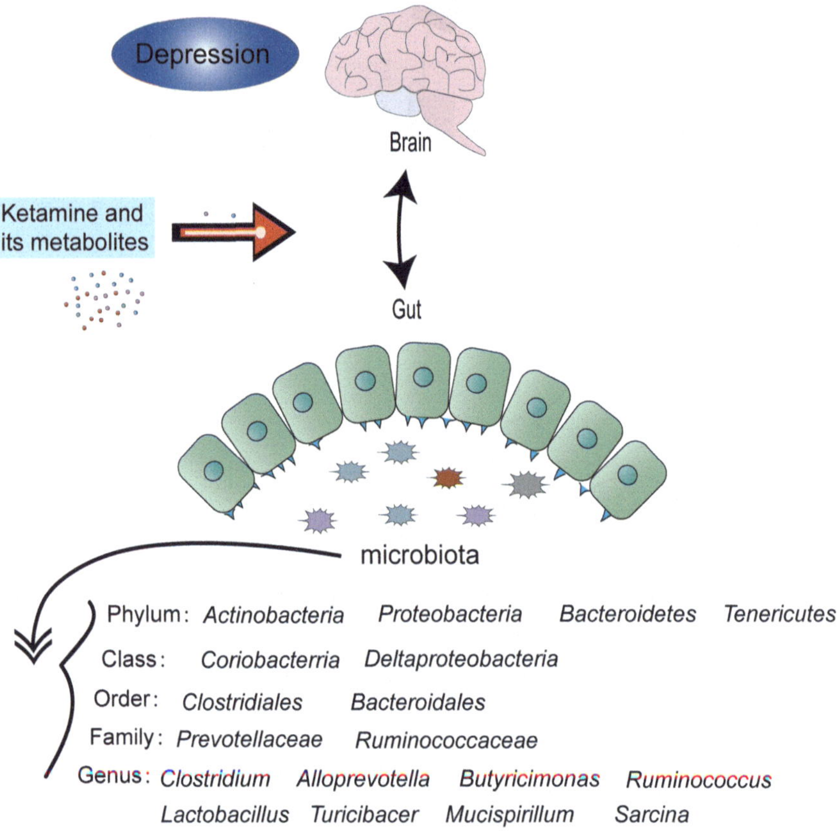

Fig. 2 Changes of gut microbiota in the antidepressant effects of ketamine. The gut microbiota involved in the antidepressant effect of ketamine are as follows: (1) Phylum level: *Actinobacteria, Proteobacteria, Bacteroidetes, Tenericutes*; (2) Class level: *Coriobacteriia, Deltaproteobacteria*; (3) Order level: *Clostridiales, Bacteroidales*; (4) Family level: *Prevotellaceae, Ruminococcaceae*; (5) Genus level: *Clostridium, Alloprevotella, Butyricimonas, Ruminococcus, Lactobacillus, Turicibacter, Mucispirillum, Sarcina*

5 Mechanisms Underlying Depression and the Antidepressant Effects of Ketamine

There is a growing evidence that the microbiota–gut–brain axis plays an important role in health and disease (Hold and Hansen 2019; Lavelle and Hill 2019). The bidirectional communication system between the brain and the gut involves the nervous, immune, metabolic, endocrine, and other mechanisms of the body (Borre et al. 2014). The gut microbiota not only regulates intestinal mechanisms but also the central nervous system by affecting brain development, stress response, anxiety, depression, and cognitive function (Mayer et al. 2014; Lee et al. 2015b). In recent years, investigation into the mechanisms underlying depression and the antidepressant effects of ketamine and how the gut microbiota might be involved in these has gained momentum, and it has been found that intestinal microbes may affect the organism's response to ketamine through inflammation, neuroendocrine imbalance, and the interference with neurotransmission (Fig. 3) (Vlainic et al. 2016).

5.1 *Inflammatory Response Mechanisms*

The bacterial metabolite lipopolysaccharide (LPS) has been found to affect the central nervous system (CNS) either directly by activating toll-like receptor 4 on microglial cells leading to the release of the inflammatory cytokines interleukin-1β (IL-1β), interleukin-6 (IL-6), and tumor necrosis factor-α (TNF-α) in the CNS or indirectly through inducing the release of inflammatory cytokines in the gastrointestinal tract (Shi et al. 2006; Kim et al. 2012). One study has found the presence of a negative correlation between *Actinobacteria, Coriobacteriia, Clostridiales*, and FST immobility time in the LPS-induced inflammatory depression mouse model and a positive correlation between *Prevotellaceae, Alloprevotella*, and FST immobility time, suggesting that a change in the levels of these microbial communities may lead to depression in mice through an inflammatory response. Moreover, ROC curve analysis indicated that *Actinobacillus* and *Actinomycetes*, but not *Firmicutes*, were potential biomarkers for the antidepressant ketamine in the inflammatory model of depression (Huang et al. 2019).

5.2 *Neuroendocrine Imbalance Mechanism*

It is well recognized that the hypothalamic–pituitary–adrenal (HPA) axis plays an important role in the stress response and in mood and functional disorders. Changes in the HPA axis have been detected in patients with various mental status and included elevated levels of plasma cortisol in patients with severe depression, increased levels of corticotropin-releasing factor in the cerebrospinal fluid, and 24-h

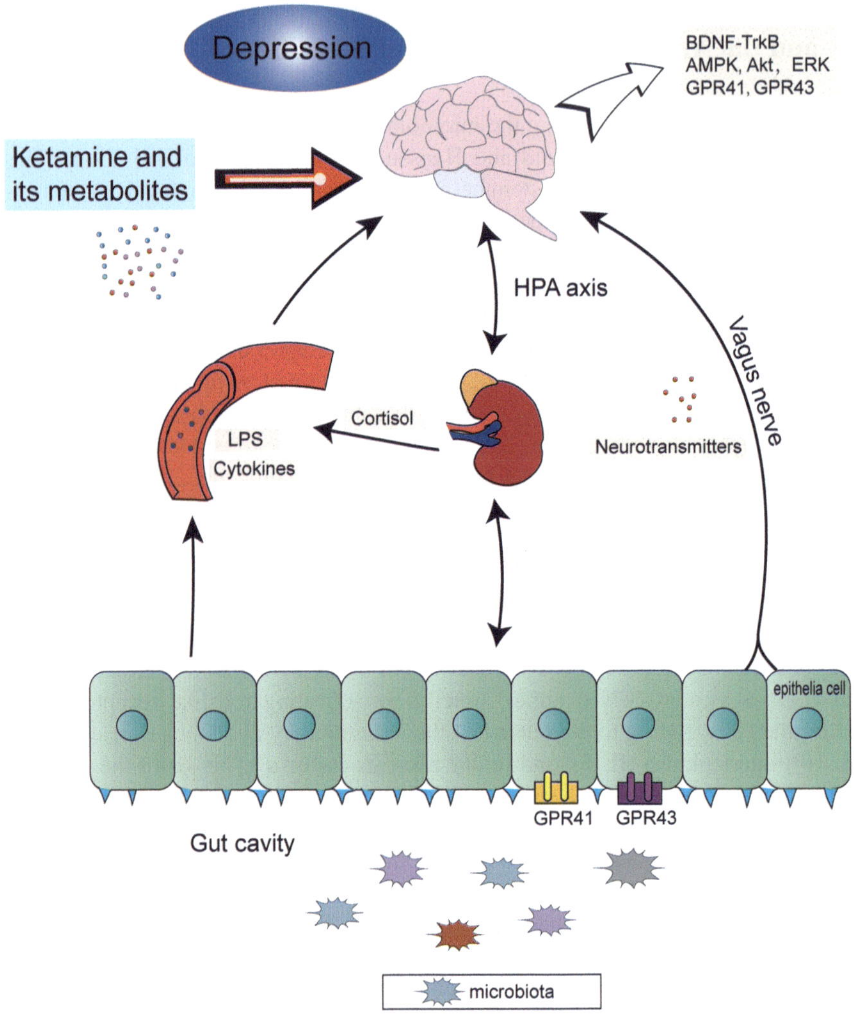

Fig. 3 Mechanisms underlying depression and antidepressant effects of ketamine. The gut microbiota may affect the antidepressant effect of ketamine through inflammatory response, neuroendocrine imbalance, and interference with neurotransmitter signals. Specifically, gut microbiota might play a pivotal role in the occurrence and development of depression by increasing the release of lipopolysaccharide (LPS) and cytokines. Furthermore, gut microbiota can activate hypothalamic–pituitary–adrenal (HPA) axis, subsequently increasing the release of cortisol into the blood, thereby promoting the antidepressant effects of ketamine. In addition, gut microbiota can also interfere with the transmission of neurotransmitters, and vagus nerve has been found to be involved in the antidepressant effects of ketamine

changes in the level of urinary 17-hydroxycorticosteroid (Carabotti et al. 2015; Barden 2004). *Lactobacillus farciminis* used as a prophylactic therapy in rats reduced the levels of the adrenal hormone and adrenocorticotropic hormones that are typically induced by stress (Ait-Belgnaoui et al. 2012). Treatment with this probiotic was also shown to prevent intestinal barrier injury, reduce circulating LPS levels, and attenuate the HPA axis stress response (Ait-Belgnaoui et al. 2012). This suggests that gut microbiota can affect the neuropsychological state of the host by regulating the HPA axis.

Gut microbiota has also been found to affect the secretion of BNDF. BDNF can be secreted by neurons in the brain and by their target cells, and its functions include promotion of nerve fiber growth and regulation of the synaptic plasticity (Jin et al. 2019). Studies have found that when the gut microbiota is modified, the mRNA and protein level of BNDF decreases, and the expression of BDNF in the hippocampus concomitantly decreases, leading to changes in the composition and function of both cortical and hippocampal neurons, which ultimately lead to depression. Therefore, gut microbiota can contribute to the mechanisms of action of antidepressant drugs by promoting BDNF expression and regulating hippocampal neuron plasticity (Maqsood and Stone 2016).

5.3 *Neurotransmitter Signaling*

Gut microbiota can promote the synthesis and release of a variety of hormones. For example, bacteria such as such as *Lactobacillus rhamnosus* and *Bifidobacterium* produce gamma-aminobutyric acid (GABA), *Bacillus* and *Saccharomyces* produce norepinephrine, *Escherichia coli* and *Enterococcus* are capable of producing 5-HT, *Bacillus* can produce dopamine, and *Lactobacillus* produces acetylcholine (Lyte 2013, 2014). These substances secreted by gut microbiota are involved in the regulation of the circadian rhythm, sexual behavior, arousal, and anxiety (Wren and Bloom 2007). Long-term treatment with *Lactobacillus rhamnosus* has been found to induce region-dependent changes in the expression of GABA receptors in the brain and to reduce stress-induced corticosterone release and anxiety and depression-like symptoms through vagus signaling (Bravo et al. 2011). Microorganisms in the gut can also affect the metabolism of tryptophan, a precursor of serotonin (5-HT). On the one hand, microbes can deplete tryptophan through the kynurenine pathway by activating indoleamine 2,3-dioxygenase, resulting in decreased levels of 5-HT and ultimately depression (Abildgaard et al. 2017b). On the other hand, tryptophan metabolism produces quinolinic acid and other neurotoxic metabolites that lead to neuronal damage (McLean et al. 2007; Weilburg 2004). Thus, there is a clear crosstalk between the substances synthesized and released by the gut microbiota and those synthesized in the host and that crosstalk can play a major role in the occurrence of central nervous system diseases.

6 Outcome of Probiotics Usage on the Antidepressant Effects of Ketamine

Traditional depression therapy typically includes the use of antidepressants and/or psychotherapy. However, there is a low recognition rate for this disease leading to be lated or non-existent medical treatment or prevention of depression, and therefore, it still presents with a high recurrence rate (Smith 2014). The use of probiotics in the field of clinical psychiatry is sparse, but it may open a new pathway for the prevention and treatment of depression as in-depth knowledge of the mechanisms for probiotic action and how they affect depression emerges. Previous studies have shown that an unhealthy diet and environmental cues can lead to a poor physical and mental state, triggering or maintaining existent depression (Tannock and Savage 1974). Probiotics can cause changes in the gut microbiota of depressed patients and significantly reduce the symptoms of depression (Liu et al. 2019).

Probiotics typically refer to a group of active microorganisms beneficial to the host through its colonization that can improve general immunity, maintain structural balance of the gut microbiota, inhibit intestinal inflammation, protect the intestinal mucosal barrier, and promote a better digestion and absorption of nutrients (Vlainic et al. 2016). Probiotics such as *Lactobacillus* and *Bifidobacteria* can noticeably activate phagocytosis by macrophages, induce the production of interferons, promote cell division, lead to the production of antibodies, and generally promote cellular immunity. Additionally, probiotics can colonize the intestine and form a natural immune barrier. Standardized mixtures of probiotics mainly include edible strains of *yeast*, *Lactobacillus acidophilus*, *Bifidobacterium*, *Clostridium butyricum*, etc. All of these have been previously shown to alleviate depression symptoms in laboratory animals (Sudo et al. 2004; Yang et al. 2017a; Liang et al. 2015).

Studies have explored the benefits of the probiotic *Bifidobacteria* in germ-free mice and found that after supplementation with infant *Bifidobacteria*, the anxiety and depression behaviors in these mice were significantly reduced. Moreover, the typically abnormal HPA axis found in germ-free mice was restored after treatment, the mRNA expression of BDNF increased, and the release of pro-inflammatory IL-6 decreased (Sudo et al. 2004). Plasma concentration of tryptophan and urinary quinolinic acid was increased in Sprague–Dawley rats that were administered with infant *Bifidobacterium* for a prolonged time, while the concentration of 5-hydroxyindoleacetic acid (5-HIAA) in the frontal cortex and 3,4-dihydroxyphenylacetic acid (DOPAC) in the amygdala was decreased, factors that typically correlate with the pathogenesis of depression. The decreased pro-inflammatory immune responses (decreased IFN-γ, TNF-α, and IL-6) and increased serotonin precursor tryptophan induced by *Bifidobacteria* therapy support the hypothesis that these bacteria can show antidepressant properties (Desbonnet et al. 2008). *Bifidobacteria* supplements may then prevent or minimize the recurrence of depression induced by either inflammation or stress (Yang et al. 2017a).

Lactic acid bacteria (LAB) is a generic term for bacteria that use fermentable carbohydrates to produce large amounts of lactic acid. LAB counteracted depression-like behavior induced by poor diet such as high-fat and low-fiber diet in rats, and an increase in white blood cell count was found in the brain tissues of depressed rats that did not receive LAB treatment (Abildgaard et al. 2017a). It has been suggested that *Lactobacillus* can exert antidepressant effects by inhibiting tryptophan-kynurenine metabolism disorder and by producing large amounts of reactive oxygen species. Further, additional supplementation with *Lactobacillus helveticus NS8* in chronic restraint-stressed rats regulates gut microbiota, improved general anxiety, depression, and memory, increased hippocampal serotonin levels and BDNF expression, and reduced cortisol and IL-10 levels (Liang et al. 2015).

Clinical exploratory experiments have shown that the use of probiotics impacts patient's anxiety and depression-like symptoms. Following *Bifidobacterium* and *Lactobacillus* supplementation for 30 days in healthy adults, these probiotics were found to reduce and alleviate psychological stress, including depression (Messaoudi et al. 2011). In another similar study, subjects were given probiotic yogurt or a combination of probiotic capsules for 6 weeks, and their mental health status was measured prospectively with an overall health questionnaire that scored depression, anxiety, and stress. Subjects receiving probiotic supplements had shown significantly lower psychological stress levels and depression scores (Mohammadi et al. 2016). Further to this, the effects of probiotic supplementation on depressive symptoms were analyzed by metabolic profiling, and measurement of serum high-sensitivity C-reactive protein (hs-CRP) and oxidative stress biomarkers was performed in patients with MDD. Following intake of probiotics for 8 weeks (*lactobacillus acidophilus*, *lactobacillus casei*, and two kinds of *Bifidobacterium*), the Beck depression scores in patients were significantly reduced with a corresponding beneficial effects in insulin, insulin resistance index, hs-CRP concentration, and glutathione levels, but not in fasting blood glucose, steady-state model evaluation of β-cell function, quantitative insulin sensitivity check index, lipid spectrum, and total antioxidant capacity levels (Akkasheh et al. 2016).

Presently, investigation of the effects of probiotics has focused mainly on infection, gastrointestinal dysfunction, and on their potential reduction of the risk of colon cancer. With the increasing usage of probiotics, their toxic and other negative side effects have started to emerge. The most common side effects of uptake of probiotics are infection, gene transfer, promotion of harmful metabolic activities, and other serious complications including mycosemia. Therefore, individualized use of probiotics that has been based on medical evidence could in future be an auxiliary method to the treatment of depression.

References

Abildgaard A, Elfving B, Hokland M et al (2017a) Probiotic treatment protects against the pro-depressant-like effect of high-fat diet in Flinders Sensitive Line rats. Brain Behav Immun 65:33–42

Abildgaard A, Elfving B, Hokland M et al (2017b) Probiotic treatment reduces depressive-like behaviour in rats independently of diet. Psychoneuroendocrinology 79:40–48

Ago Y, Tanabe W, Higuchi M et al (2019) (R)-ketamine induces a greater increase in prefrontal 5-HT release than (S)-ketamine and ketamine metabolites via an AMPA receptor-independent mechanism. Int J Neuropsychopharmacol 22(10):665–674. https://doi.org/10.1093/ijnp/pyz041

Ait-Belgnaoui A, Durand H, Cartier C et al (2012) Prevention of gut leakiness by a probiotic treatment leads to attenuated HPA response to an acute psychological stress in rats. Psychoneuroendocrinology 37(11):1885–1895

Akkasheh G, Kashani-Poor Z, Tajabadi-Ebrahimi M et al (2016) Clinical and metabolic response to probiotic administration in patients with major depressive disorder: a randomized, double-blind, placebo-controlled trial. Nutrition 32(3):315–320

Barden N (2004) Implication of the hypothalamic-pituitary-adrenal axis in the physiopathology of depression. J Psychiatry Neurosci 29(3):185–193

Bartoli F, Riboldi I, Crocamo C et al (2017) Ketamine as a rapid-acting agent for suicidal ideation: a meta-analysis. Neurosci Biobehav Rev 77:232–236

Berman RM, Cappiello A, Anand A et al (2000) Antidepressant effects of ketamine in depressed patients. Biol Psychiatry 47(4):351–354

Borre YE, Moloney RD, Clarke G et al (2014) The impact of microbiota on brain and behavior: mechanisms and therapeutic potential. Adv Exp Med Biol 817:373–403

Bortolozzi A, Celada P, Artigas F (2014) Novel therapeutic strategies in major depression: focus on RNAi and ketamine. Curr Pharm Des 20(23):3848–3860

Bravo JA, Forsythe P, Chew MV et al (2011) Ingestion of Lactobacillus strain regulates emotional behavior and central GABA receptor expression in a mouse via the vagus nerve. Proc Natl Acad Sci U S A 108(38):16050–16055

Carabotti M, Scirocco A, Maselli MA et al (2015) The gut-brain axis: interactions between enteric microbiota, central and enteric nervous systems. Ann Gastroenterol 28(2):203–209

Chou D, Peng HY, Lin TB et al (2018) (2R,6R)-hydroxynorketamine rescues chronic stress-induced depression-like behavior through its actions in the midbrain periaqueductal gray. Neuropharmacology 139:1–12

Clark-Raymond A, Halaris A (2013) VEGF and depression: a comprehensive assessment of clinical data. J Psychiatr Res 47(8):1080–1087

Desbonnet L, Garrett L, Clarke G et al (2008) The probiotic Bifidobacteria infantis: an assessment of potential antidepressant properties in the rat. J Psychiatr Res 43(2):164–174

Deyama S, Bang E, Wohleb ES et al (2019) Role of neuronal VEGF signaling in the prefrontal cortex in the rapid antidepressant effects of ketamine. Am J Psychiatry 176(5):388–400

Dinan TG, Cryan JF (2019) Gut microbes and depression: still waiting for godot. Brain Behav Immun 79:1–2

Fukumoto K, Toki H, Iijima M et al (2017) Antidepressant potential of (*R*)-ketamine in rodent models: comparison with (*S*)-ketamine. J Pharmacol Exp Ther 361(1):9–16

Getachew B, Aubee JI, Schottenfeld RS et al (2018) Ketamine interactions with gut-microbiota in rats: relevance to its antidepressant and anti-inflammatory properties. BMC Microbiol 18(1):222

Goitsuka R, Hirota Y, Hasegawa A et al (1987) Release of interleukin 1 from peritoneal exudate cells of cats with feline infectious peritonitis. Nihon Juigaku Zasshi 49(5):811–818

Grunebaum MF, Galfalvy HC, Choo TH et al (2018) Ketamine for rapid reduction of suicidal thoughts in major depression: a midazolam-controlled randomized clinical trial. Am J Psychiatry 175(4):327–335

Hashimoto K (2019) Rapid-acting antidepressant ketamine, its metabolites and other candidates: a historical overview and future perspective. Psychiatry Clin Neurosci 73(10):613–627. https://doi.org/10.1111/pcn.12902

Hoban AE, Moloney RD, Golubeva AV et al (2016) Behavioural and neurochemical consequences of chronic gut microbiota depletion during adulthood in the rat. Neuroscience 339:463–477

Hold GL, Hansen R (2019) impact of the gastrointestinal microbiome in health and disease: co-evolution with the host immune system. Curr Top Microbiol Immunol 421:303–318

Huang N, Hua D, Zhan G et al (2019) Role of actinobacteria and coriobacteriia in the antidepressant effects of ketamine in an inflammation model of depression. Pharmacol Biochem Behav 176:93–100

Jiang H, Ling Z, Zhang Y et al (2015) Altered fecal microbiota composition in patients with major depressive disorder. Brain Behav Immun 48:186–194

Jin Y, Sun LH, Yang W et al (2019) The role of BDNF in the neuroimmune axis regulation of mood disorders. Front Neurol 10:515

Kim KA, Gu W, Lee IA et al (2012) High fat diet-induced gut microbiota exacerbates inflammation and obesity in mice via the TLR4 signaling pathway. PLoS One 7(10):e47713

Lavelle A, Hill C (2019) Gut microbiome in health and disease: emerging diagnostic opportunities. Gastroenterol Clin North Am 48(2):221–235

Lee EE, Della Selva MP, Liu A et al (2015a) Ketamine as a novel treatment for major depressive disorder and bipolar depression: a systematic review and quantitative meta-analysis. Gen Hosp Psychiatry 37(2):178–184

Lee SP, Sung IK, Kim JH et al (2015b) The effect of emotional stress and depression on the prevalence of digestive diseases. J Neurogastroenterol Motil 21(2):273–282

Liang S, Wang T, Hu X et al (2015) Administration of Lactobacillus helveticus NS8 improves behavioral, cognitive, and biochemical aberrations caused by chronic restraint stress. Neuroscience 310:561–577

Liu RT, Walsh RFL, Sheehan AE (2019) Prebiotics and probiotics for depression and anxiety: a systematic review and meta-analysis of controlled clinical trials. Neurosci Biobehav Rev 102:13–23

Lyte M (2013) Microbial endocrinology in the microbiome-gut-brain axis: how bacterial production and utilization of neurochemicals influence behavior. PLoS Pathog 9(11):e1003726

Lyte M (2014) Microbial endocrinology and the microbiota-gut-brain axis. Adv Exp Med Biol 817:3–24

Malhi GS, Mann JJ (2018) Depression. Lancet 392(10161):2299–2312

Maqsood R, Stone TW (2016) The gut-brain axis, BDNF, NMDA and CNS disorders. Neurochem Res 41(11):2819–2835

Mastrodonato A, Martinez R, Pavlova IP et al (2018) Ventral CA3 activation mediates prophylactic ketamine efficacy against stress-induced depressive-like behavior. Biol Psychiatry 84(11):846–856

Mayer EA, Padua D, Tillisch K (2014) Altered brain-gut axis in autism: comorbidity or causative mechanisms? Bioessays 36(10):933–939

McGirr A, Berlim MT, Bond DJ et al (2015) A systematic review and meta-analysis of randomized, double-blind, placebo-controlled trials of ketamine in the rapid treatment of major depressive episodes. Psychol Med 45(4):693–704

McLean PG, Borman RA, Lee K (2007) 5-HT in the enteric nervous system: gut function and neuropharmacology. Trends Neurosci 30(1):9–13

Messaoudi M, Lalonde R, Violle N et al (2011) Assessment of psychotropic-like properties of a probiotic formulation (Lactobacillus helveticus R0052 and Bifidobacterium longum R0175) in rats and human subjects. Br J Nutr 105(5):755–764

Mohammadi AA, Jazayeri S, Khosravi-Darani K et al (2016) The effects of probiotics on mental health and hypothalamic-pituitary-adrenal axis: a randomized, double-blind, placebo-controlled trial in petrochemical workers. Nutr Neurosci 19(9):387–395

Murrough JW, Iosifescu DV, Chang LC et al (2013) Antidepressant efficacy of ketamine in treatment-resistant major depression: a two-site randomized controlled trial. Am J Psychiatry 170(10):1134–1142

Murrough JW, Soleimani L, DeWilde KE et al (2015) Ketamine for rapid reduction of suicidal ideation: a randomized controlled trial. Psychol Med 45(16):3571–3580

Pennisi E (2019) Gut bacteria linked to mental well-being and depression. Science 363(6427):569

Peyrovian B, Rosenblat JD, Pan Z et al (2019) The glycine site of NMDA receptors: A target for cognitive enhancement in psychiatric disorders. Prog Neuropsychopharmacol Biol Psychiatry 92:387–404

Price RB, Nock MK, Charney DS et al (2009) Effects of intravenous ketamine on explicit and implicit measures of suicidality in treatment-resistant depression. Biol Psychiatry 66(5):522–526

Qin J, Li R, Raes J et al (2010) A human gut microbial gene catalogue established by metagenomic sequencing. Nature 464(7285):59–65

Qu Y, Yang C, Ren Q et al (2017) Comparison of (R)-ketamine and lanicemine on depression-like phenotype and abnormal composition of gut microbiota in a social defeat stress model. Sci Rep 7(1):15725

Reardon S (2019) Antidepressant based on party drug gets backing from FDA advisory group. https://www.nature.com/articles/d41586-019-00559-2. Accessed 13 Feb 2019

Shi H, Kokoeva MV, Inouye K et al (2006) TLR4 links innate immunity and fatty acid-induced insulin resistance. J Clin Invest 116(11):3015–3025

Skolnick P, Layer RT, Popik P et al (1996) Adaptation of N-methyl-D-aspartate (NMDA) receptors following antidepressant treatment: implications for the pharmacotherapy of depression. Pharmacopsychiatry 29(1):23–26

Smith K (2014) Mental health: a world of depression. Nature 515(7526):181

Stower H (2019) Depression linked to the microbiome. Nat Med 25(3):358

Sudo N, Chida Y, Aiba Y et al (2004) Postnatal microbial colonization programs the hypothalamic-pituitary-adrenal system for stress response in mice. J Physiol 558(Pt 1):263–275

Sun HL, Zhou ZQ, Zhang GF et al (2016) Role of hippocampal p11 in the sustained antidepressant effect of ketamine in the chronic unpredictable mild stress model. Transl Psychiatry 6:e741

Tannock GW, Savage DC (1974) Influences of dietary and environmental stress on microbial populations in the murine gastrointestinal tract. Infect Immun 9(3):591–598

Trivedi MH, Rush AJ, Wisniewski SR et al (2006) Evaluation of outcomes with citalopram for depression using measurement-based care in STAR∗D: implications for clinical practice. Am J Psychiatry 163(1):28–40

U.S. Food and Drug Administration (2019) FDA approves new nasal spray medication for treatment-resistant depression; available only at a certified doctor's office or clinic. https://www.fda.gov/news-events/press-announcements/fda-approves-new-nasal-spray-medication-treatment-resistant-depression-available-only-certified. Accessed 6 Mar 2019

Vlainic JV, Suran J, Vlainic T et al (2016) Probiotics as an adjuvant therapy in major depressive disorder. Curr Neuropharmacol 14(8):952–958

Warden D, Rush AJ, Trivedi MH et al (2007) The STAR∗D Project results: a comprehensive review of findings. Curr Psychiatry Rep 9(6):449–459

Weilburg JB (2004) An overview of SSRI and SNRI therapies for depression. Manag Care 13(6 Suppl Depression):25–33

Williams NR, Heifets BD, Blasey C et al (2018) Attenuation of antidepressant effects of ketamine by opioid receptor antagonism. Am J Psychiatry 175(12):1205–1215

World Health Organization (2017) Depression: let's talk. https://www.who.int/mental_health/management/depression/en/. Accessed 7 Apr 2017

Wren AM, Bloom SR (2007) Gut hormones and appetite control. Gastroenterology 132(6):2116–2130

Xu Y, Hackett M, Carter G et al (2016) Effects of low-dose and very low-dose ketamine among patients with major depression: a systematic review and meta-analysis. Int J Neuropsychopharmacol 19(4):pyv124

Yang J, Yu J (2018) The association of diet, gut microbiota and colorectal cancer: what we eat may imply what we get. Protein Cell 9(5):474–487

Yang C, Shirayama Y, Zhang JC et al (2015) *R*-ketamine: a rapid-onset and sustained antidepressant without psychotomimetic side effects. Transl Psychiatry e632:5

Yang C, Fujita Y, Ren Q et al (2017a) Bifidobacterium in the gut microbiota confer resilience to chronic social defeat stress in mice. Sci Rep 7:45942

Yang C, Qu Y, Fujita Y et al (2017b) Possible role of the gut microbiota-brain axis in the antidepressant effects of (R)-ketamine in a social defeat stress model. Transl Psychiatry 7(12):1294

Yang C, Kobayashi S, Nakao K et al (2018a) AMPA receptor activation-independent antidepressant actions of ketamine metabolite (S)-norketamine. Biol Psychiatry 84(8):591–600
Yang C, Ren Q, Qu Y et al (2018b) Mechanistic target of rapamycin-independent antidepressant effects of (R)-ketamine in a social defeat stress model. Biol Psychiatry 83(1):18–28
Zarate CA Jr, Singh JB, Carlson PJ et al (2006) A randomized trial of an N-methyl-D-aspartate antagonist in treatment-resistant major depression. Arch Gen Psychiatry 63(8):856–864
Zarate CA Jr, Brutsche NE, Ibrahim L et al (2012) Replication of ketamine's antidepressant efficacy in bipolar depression: a randomized controlled add-on trial. Biol Psychiatry 71(11):939–946
Zhang K, Hashimoto K (2019) Lack of opioid system in the antidepressant actions of ketamine. Biol Psychiatry 85(6):e25–e27
Zhang JC, Li SX, Hashimoto K (2014) R (-)-ketamine shows greater potency and longer lasting antidepressant effects than S (+)-ketamine. Pharmacol Biochem Behav 116:137–141
Zhang K, Xu T, Yuan Z et al (2016) Essential roles of AMPA receptor GluA1 phosphorylation and presynaptic HCN channels in fast-acting antidepressant responses of ketamine. Sci Signal 9(458):ra123
Zhang JC, Yao W, Dong C et al (2017) Blockade of interleukin-6 receptor in the periphery promotes rapid and sustained antidepressant actions: a possible role of gut-microbiota-brain axis. Transl Psychiatry 7(5):e1138
Zheng P, Zeng B, Zhou C et al (2016) Gut microbiome remodeling induces depressive-like behaviors through a pathway mediated by the host's metabolism. Mol Psychiatry 21(6):786–796

Effects of Ketamine on Pain and Depression Comorbidity

Xiao-Hui Tang and Jian-Jun Yang

Abstract Major depressive disorder is one of the leading causes of disability in the world, with an incidence rate of 16%, causing a heavy social and economic burden. Moreover, pain and depression often coexist, and the factors that need to be considered to understand this coexistence are more complex. A deeper understanding of the mechanisms behind the coexistence of pain and depression is likely to find a more effective way to treat depression in the context of pain. In this chapter, we outline the prevalence of pain and depression comorbidities and the shared neurobiological mechanisms between pain and depression. Additionally, we discuss animal models commonly used in the current study to simulate comorbidities of pain and depression. Finally, we also present some current therapies based on an understanding of the neurobiological changes that occur in pain and depression comorbidities, as well as ketamine as a promising approach in the treatment of pain and depression comorbidities.

Keywords Pain · Depression · Ketamine animal models · Glutamate signaling

1 Relationship Between Pain and Depression

Depression and pain are two highly prevalent and deleterious disorders (Bair et al. 2003), which make detrimental effects on patient health and cause a significant economic burden to society (Sobocki et al. 2006). There is growing evidence suggests that these two situations commonly coexist and often intensify each other and may have overlapping symptoms (Li 2015). According to reports, the prevalence rates of chronic pain in patients with depression is about 59.1%, higher than the general population (Aguera-Ortiz et al. 2011). According to the Symptom Checklist (SCL) scale, people with pain symptoms (e.g., back pain, headache, abdominal pain, chest pain, and facial pain) have a higher depression score than those without pain symptoms (Von Korff et al. 1988). Meanwhile, the average prevalence rates of

X.-H. Tang · J.-J. Yang (✉)
Department of Anesthesiology, Pain and Perioperative Medicine, The First Affiliated Hospital of Zhengzhou University, Zhengzhou, China

K. Hashimoto et al. (eds.), *Ketamine*,
https://doi.org/10.1007/978-981-15-2902-3_9

major depression in patients identified as suffering from pain ranged from 18 to 85%, depending on the study diagnostic criteria (Bair et al. 2003). A 10-year follow-up found that depressive symptoms can be used as predictors of low back pain and neck–shoulder pain (Leino and Magni 1993). Therefore, the current consensus is that pain and depression often present together.

Pain is defined as "an unpleasant subjective feeling and emotional experience associated with actual or potential tissue damage," which is an interaction of psychological, emotional, behavioral, and social factors. Among them, chronic pain means pain persists or recurs for more than 3 months. Pain is a complex that combines sensory, cognitive, and emotional components. The emotional component is the most important, and depression is the outstanding performance of the emotional component. Studies suggested that pain and depression were highly intertwined and may co-exacerbate physical and psychological symptoms. Chronic painful physical condition (CPPC) is a risk indicator for developing major depressive disorder (MDD), which exacerbates the severity of depression symptoms, prolongs the duration of depressive episodes, and increases recurrence rates (Ohayon 2004). Meanwhile, Means-Christensen et al. observed that patients with pain symptoms had decreased mental health function and a higher score for depression severity, and there is a 2.5–10 times increase in depression in patients with pain (Means-Christensen et al. 2008). Furthermore, in the postoperative period, the severity of depressive symptoms is closely related to the intensity of pain, as the depression score increases with increasing pain (Carr et al. 2005). The onset of depression is associated with an increasing number of pain locations ($P < 0.001$) and higher severity of pain ($P < 0.001$) (Gerrits et al. 2014).

In terms of the temporal relationship between pain and depression, we mainly discuss the coexistence of such a disease in which pain is in the front and depression is behind. On the one hand, pain precedes and subsequently triggers depression-like symptoms, a view that is consistently supported by clinical and animal experimental data. On the other hand, the state of depression precedes and subsequently changes the sensitivity of the pain, and the results obtained from the study of this view are less consistent. In general, it must be pointed out that in the study of the temporal relation between pain and depression, Hilderink et al. concluded that since pain precedes the onset of depression, strategies to prevent depression in patients with chronic pain are necessary. In contrast, no effect of depression on subsequent pain development was found when adjusting for covariates (Hilderink et al. 2012). In summary, these findings suggest that we must be aware that pain plays an important role in depression and that a better understanding of pain and depression comorbidity can be helpful for the treatment of depression.

2 Mechanisms of Pain and Depression Comorbidity

2.1 *Why Pain and Depression Are Associated?*

Structure determines function. To understand why pain and depression are related, it is easy to think about whether there is a connection or overlap in the brain regions affected by the two. Importantly, functional imaging results confirm this (Hooten 2016). Pain is a complex sensory and emotional experience, and multiple pathways in the central nervous system (CNS) are involved in the management of pain. Human brain imaging studies have revealed that the brain areas most commonly activated include the primary and secondary somatosensory, insular, anterior cingulate, and prefrontal cortices (S1, S2, IC, ACC, PFC) and thalamus (Th). Among them, ACC, nucleus accumbens (NAc), and amygdala are thought to be related to the emotional component of pain (Apkarian et al. 2005; Bushnell et al. 2013). Similarly, the most frequently dysfunctional brain regions of depression include PFC, ACC, NAc, hippocampus, and amygdala (Fig. 1). Interestingly, the brain areas affected by chronic pain and depression are very similar (Russo and Nestler 2013). In healthy subjects, patterns of brain activity caused by emotional distress (using experimental means of social exclusion that leads to depression) are similar to patterns of brain activity caused by experimental pain, including somatosensory and active areas (Eisenberger et al. 2003).

Preclinical studies support the structural relevance of pain and depression. A recent study showed that NAc is participated in both pain relief and reward circuitry

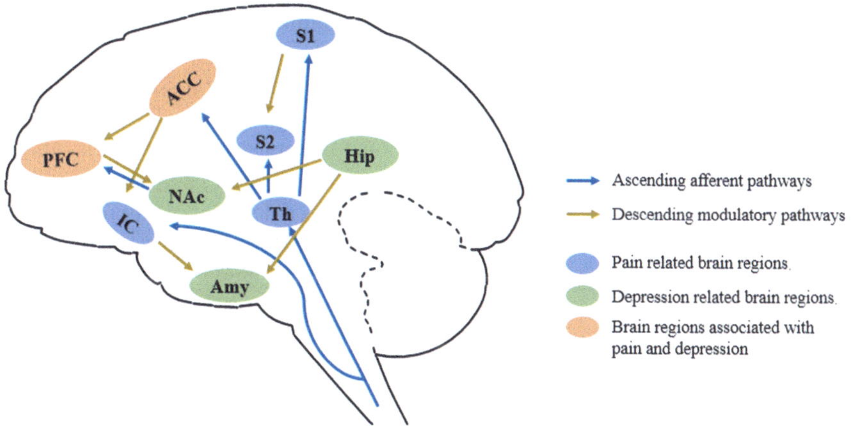

Fig. 1 Brain regions associated with pain and depression comorbidity. Afferent nociceptive information (pain) transmitted from the spinal cord to the brain, activating a series of brain regions, including the primary and secondary somatosensory, insular, anterior cingulate, and prefrontal cortices (S1, S2, IC, ACC, PFC) and thalamus (Th), whereas depression activates brain regions involving PFC, ACC, NAc, hippocampus, and amygdala. There is an interaction between the brain regions involved in the ascending (shown in blue) afferent and descending (shown in yellow) modulatory pathways

(Baliki et al. 2013). Moreover, in a mouse model of SNI, the firing rate and bursting activity in the ACC increased when depressive-like behaviors emerged as a consequence of neuropathic pain, and optogenetic inhibition of the ACC hyperactivity alleviated the depressive-like behaviors (Sellmeijer et al. 2018). At the same time, clinical studies also support the structural relevance of pain and depression. Functional imaging results have showed that, accompanying the progress that acute pain transform chronic pain, the brain associated with emotion and reward was activated (Bigman et al. 2017). In addition, compared with healthy subjects, the emotionally relevant areas of depressed patients were transferred to the dorsal anterior island, which is the area associated with physical pain (Mutschler et al. 2012). All these studies noted that there are common brain changes (structure level) in pain and depression comorbidity (function level).

2.2 *What Is Involved in Pain and Depression Comorbidity?*

The neurobiological mechanisms of pain and depression comorbidity have not been extensively studied; on the one hand, the simultaneous occurrence of these two conditions may indicate that there is some common biological mechanism behind it (Cocksedge et al. 2014), and on the other hand, the difference in the treatment effect of pain and depression may indicate that the mechanism behind it is also different. Thus, it is clearly that the mechanism is multifaceted.

2.2.1 Animal Models of Pain and Depression Comorbidity

Although imaging studies suggest a link between pain and depression at the anatomical level, further mechanistic studies need to be explored through rational laboratory animal models. In a widely accepted model of depression, pain often acts as a source of stress to induce symptoms of depression; therefore, a model for establishing pain and depression comorbidities can be expected. The most representative of them is chronic neuropathic pain models, which simulates the depression caused by chronic pain in the population. Clinical studies have shown that patients with various pain conditions have an increased susceptibility to depression. Almost all preclinical studies on depression in the context of pain use models associated with sciatic nerve manipulation, using nerve ligation or injury. Among them, chronic constriction injury (CCI), sciatic nerve ligation model (SNL), and spared nerve injury (SNI) model are commonly used in experiments (Decosterd and Woolf 2000). Anhedonia and behavioral despair are the two most important features of depression. The sucrose preference test (SPT) and the forced swim test (FST) were used to assess the anhedonia and behavioral despair, respectively. Many studies have confirmed that both SNI and SNL can induce anhedonia as well as behavioral despair (Wang et al. 2011; Goffer et al. 2013; Suzuki et al. 2007). In a rat model of CCI, during day 14 to day 21 after CCI, the immobility (depression-like behavior) time of FST increased (Fukuhara et al. 2012). In a mice model of SNL, 15 and 30 days after ligation, mice developed depression-related behavior, which was demonstrated by an increase in immobility time of FST (Suzuki et al. 2007). In a rat model of SNI, rat developed depression-related behaviors, as evidenced by reduced sucrose preference in SPT and increased immobility time in FST (Goffer et al. 2013). In addition to the chronic neuropathic pain model described above, inflammatory model of pain, like complete Freund's adjuvant (CFA) model has also been shown to induce depression-like symptoms (Shi et al. 2010). All of these models, which have been used and validated, serve as a valuable tool for the study of the comorbidity between pain and depression. While, in models of pain-induced depression, depression-like behavior occurs later than pain symptoms, and further study and understanding of this behavioral synchronization is necessary for a comprehensive understanding of the comorbid model. Furthermore, because the current type of pain is too single, the commonly used models are derived from neuropathic pain. Future research should include more types of pain to establish a depression model, in order to better simulate the clinical situation.

2.3 Molecular Mechanisms That Underlie the Pain and Depression Comorbidity

2.3.1 Glutamate Signaling

Glutamate is the major excitatory neurotransmitter in the nervous system and functions by binding to receptors, among which the common ionotropic glutamate receptors are α-amino-3-hydroxy-5-methyl-4-isoxazolepropionic acid (AMPA) and *N*-methyl-D-aspartate (NMDA) receptors. Glutamatergic system as a final common pathway for the novel effective antidepressant indicated its important role in psychiatric pathology (Sanacora et al. 2012). In the rodent models of depression, the levels of AMPA receptors subunit GluA1 in the PFC were reduced (Chandran et al. 2013). Corresponding antidepressants can upregulate the expression of GluA1 (Li et al. 2010). Similarly, glutamate signaling is involved in the process of pain as well. In the rat CCI model, the phosphorylation level of the NMDA receptor type 1 (NR1) subunit in the hippocampus was reduced compared to the control group (Li et al. 2014). In the SNI model for chronic pain, the levels of vesicular glutamate transporter (VGLUT) 1 and 3 levels in the NAc were decreased (Tukey et al. 2013). It is reasonable to speculate that glutamatergic signaling is also participated in the pain and depression comorbidity, and some studies have confirmed this. In a rat model of chronic neuropathic pain, the level of GluA1 in NAc was increased and the calcium-permeable AMPA receptors (CPARs) was formed. Pharmacologic blockade of CPARs increases pain-related depression-like behaviors, while enhancement of AMPA receptor function decreases pain-induced depression-like behaviors (Goffer et al. 2013). Furthermore, in the SNI model, ketamine, which has been shown to enhance glutamate signaling can treat depression-like behaviors caused by pain

(Wang et al. 2011). Meanwhile, injection of AMPAkines (augment AMPA receptor function) produce analgesic effect in rat models of neuropathic and inflammatory pain (Li et al. 2014). In patients with irritable bowel syndrome (IBS), increased pain sensitivity can be blocked by an NMDA receptor antagonist (dextromethorphan), suggesting that NMDA receptor mechanisms may be involved in the treatment of pain (Verne et al. 2012). In summary, these studies indicate that glutamate signaling plays a role in the formation and treatment of pain and depression comorbidities.

2.3.2 Serotonin (5-HT), Dopamine (DA), and Norepinephrine (NE) Neurotransmitters

Decreased serotonergic activity was recognized as one of the causes of depression, and classical antidepressants also work by directly or indirectly increasing the activity of the 5-HT system (Albert et al. 2014). In depressed patients with suicidal behavior, the level of 5-HT in the cerebrospinal fluid was reduced (Mann and Malone 1997). At the same time, in patients with fibromyalgia, the maximal uptake rate of specific serotonin transporter (SERT) was decreased in the platelets (Bazzichi et al. 2006), and the level of 5-HT in the cerebrospinal fluid was also decreased. Moreover, the 5-HT system was associated with hyperalgesia and central sensitization, which is caused by the descending 5-HT drive from the rostral ventromedial medulla and the involvement of 5-HT receptors (Okubo et al. 2013). Experimental studies closer to comorbidities have shown that in patients with both pain and depression, the level of indoleamine 2,3-dioxygenase 1 (IDO1), a rate-limiting enzyme associated with a precursor of serotonin, in plasma was elevated. Furthermore, results show that both nociceptive behavior and depressive behavior were attenuated when IDO1 gene knockout or pharmacological inhibition of IDO1 activity (Kim et al. 2012). These studies provide evidences that 5-HT signaling is associated with pain and depression comorbidity, which may be related to the metabolic pathway of 5-HT signaling.

Clinical and basic research findings suggest that NE plays a role in the pathophysiology of depression. In depressed patients, the level of NE in the cerebrospinal fluid was reduced. NE depletion can reproduce the clinical symptoms of depression, and antidepressants can reverse this change (Delgado 2000). Moreover, the levels of tyrosine hydroxylase (rate-limiting enzyme in NE biosynthesis) and amounts of binding to NE receptors were increased in depressed suicides compared with the control group (Ordway 1997). At the same time, chronic stress has been found to alter the transmission of NE and is further exacerbated by comorbid chronic pain (Bravo et al. 2014). In patients with complex regional pain syndrome, neuropathic pain, and cancer pain, the a2-adrenergic receptor agonist clonidine has been reported to have an analgesic effect (Ackerman et al. 2003). Experimental studies closer to comorbidities have shown that, in a rat model of CCI, pain leads to depressive-like behaviors, and the onset of these behaviors are consistent with changes in the NE system (Alba-Delgado et al. 2013). It should be noted that this change is different, that is, the changes in the NE system caused by different types of pain, such as neuropathic pain or inflammatory pain, are not exactly the same. More research is needed in the future to clarify the specific roles played by different pains.

DA is considered to be involved in the pathophysiology and treatment of depression (Moriam and Sobhani 2013). In healthy individuals and depressed patients, a lower dopamine genetic risk score (indicating lower dopaminergic neurotransmission) can predict a higher level of depression (Pearson-Fuhrhop et al. 2014). At the same time, studies have shown that migraine was characterized by chronic dopaminergic deficiency (Barbanti et al. 2013). Experimental studies closer to comorbidities have shown that, in a rat model of SNI, the burst firing of DA cells in the SNI group is enhanced compared to that in sham-operated animals (Sagheddu et al. 2015). 5-HT, DA, and NE are all monoamine neurotransmitters. Recently, triple reuptake inhibitors that simultaneously inhibit 5-HT, NE, and DA reuptake reversed the neuropathic hypersensitivity in a mouse model of neuropathic pain (Hache et al. 2015). These results accentuate the role of monoamine neurotransmitters in the pain and depression comorbidity.

2.3.3 Brain-Derived Neurotrophic Factor (BDNF) Signaling

BDNF is a widely expressed neurotrophic protein in the peripheral and central nervous systems that is involved in the regulation of neuronal survival and differentiation and is critical for the regulation of synaptic plasticity. Converging lines of evidence implicate the BDNF in the pathophysiology of major depression. Clinical experiments have shown that, in depressed patients, the level of BDNF in serum was decreased (Karege et al. 2002). Other studies have shown that, after antidepressant treatment, the improvement of clinical symptoms in drug-free depressed patients is parallel with the rise of plasma BDNF (Piccinni et al. 2008). At the same time, in patients suffering from pain (headaches), the platelet levels of BDNF were decreased (Blandini et al. 2006) (Table 1). Furthermore, BDNF is essential for central sensitization, which is a major component for persistent pain states, and pharmacological antagonism of BDNF can treat persistent inflammatory pain (Kerr et al. 1999). More convincing experiments have shown that BDNF conditionally knockout animals have a reduced baseline value of thermal pain and show that BDNF plays an important role in regulating inflammatory pain thresholds (Zhao et al. 2006).

Unlike the results observed in clinical experiments, the results obtained in preclinical experiments have some regional regulatory characteristics that need to be further clarified. In spinal cord dorsal horn, the expression of BDNF was increased in a rat model of CFA and reversed by anti-inflammatory drug (Duric and McCarson 2006a). In the dorsal root ganglion, the expression of BDNF was increased in a mouse model of SNI and reversed by swimming exercise (Almeida et al. 2015). Differentially, in the hippocampus, the expression of BDNF is reduced in a rat model of CFA and can be blocked by antidepressants (Duric and McCarson 2006a). In a rat model of SNL, the level of BDNF mRNA in the hippocampus decreased 7–14 days after SNL, and BDNF treatment restored hyperalgesia behavior (Tateiwa et al. 2018). This different change in BDNF at the spinal and brain levels suggests

Table 1 Similar biomarkers between depression and pain

Biomarker	Depression	Pain
Glutamate	(1) Reduced GluA1 in the PFC	(1) Reduced NR1 subunit in the hippocampus
	(2) Reduced GluA1 in the NAc	(2) Reduced VGLUT 1 and 3 in the NAc
		(3) Reduced GluA1 in the NAc
Serotonin	(1) Reduced 5-HT in the CSF	(1) Reduced SERT in the platelets
	(2) Elevated IDO1 in the plasma	(2) Reduced 5-HT in the CSF
		(3) Elevated IDO1 in the plasma
Norepinephrine	(1) Reduced NE in the CSF	(1) Reduced TH in the locus coeruleus
	(2) Reduced TH in the locus coeruleus	
Dopamine	(1) Reduced DA neurotransmission	(1) Reduced DA in the plasma
	(2) Reduced DA cells burst firing	(2) Reduced DA cells burst firing
	(1) Reduced BDNF in serum	(1) Reduced BDNF in platelet
BDNF	(2) Reduced BDNF in plasma	(2) Elevated BDNF in the spinal cord
	(3) Elevated BDNF in the spinal cord	(3) Reduced BDNF in the hippocampus
	(4) Reduced BDNF in the hippocampus	

PFC prefrontal cortices, *NAc* nucleus accumbens, *VGLUT* vesicular glutamate transporter, *SERT* specific serotonin transporter, *IDO1* indoleamine 2,3-dioxygenase 1, *CSF* cerebrospinal fluid, *BDNF* brain-derived neurotrophic factor

that further experiments are needed to determine what role the peripheral and central nervous systems play in a coexisting disease, and what are the connections and differences between them.

Experimental studies closer to comorbidities have shown that in inflammatory model of pain, the levels of BDNF mRNAs in the hippocampus were reduced, which is also related to the pathophysiology of depression (Duric and McCarson 2006b). Moreover, in a rat model of CCI, the lack of BDNF is involved in depression-like behaviors during chronic pain, while enhanced BDNF can alleviate chronic pain and depression-like behaviors (Fukuhara et al. 2012). In summary, the evidence leads us to believe that BDNF signaling was involved in the comorbidities of pain and depression.

3 Treatments of Pain and Depression Comorbidity

In the treatment of pain and depression comorbidity, it must be recognized that the presence of pain may increase the resistance of depression to treatment. Depressed patients with severe pain were less likely to achieve remission and partial response than those without pain at baseline (DeVeaugh-Geiss et al. 2010). Therefore, it is more logical to choose a drug to treat pain and depression comorbidity to promote neurotransmitters that are commonly altered in both diseases.

3.1 Serotonin-Norepinephrine Reuptake Inhibitors (SNRIs)

SNRIs works by dually inhibiting the reuptake of serotonin and norepinephrine, which has been proven to be a safe, tolerable, and effective antidepressant, as well as significantly reducing pain symptoms such as headaches and back pain (Arnold 2007). The analgesic effects of SNRIs are also evident in the treatment of depression, echoing the hypothesis that serotonin and norepinephrine systems are the common biochemical basis for pain and depression. Among them, many studies have focused on the possibility that duloxetine can treat both pain and depression (a SNRIs). In patients concurrent with recurrent MDD and arthritis pain, duloxetine not only significantly reduced the severity of MDD symptom but also yielded a significant reduction in pain symptoms (Wohlreich et al. 2009). This result was supported by Brannan et al. study, which observed that depressed patients with associated painful physical symptoms produce an effective response to duloxetine (60 mg once daily) (Brannan et al. 2005). More specific studies have shown that among depressed patients grouped according to the physical symptoms of pain, depressed patients with pain symptoms have a higher rate of remission after treatment with duloxetine than depressed patients without pain symptoms (Sekine et al. 2016). One possible mechanism is that the transmitter system affected by SNRIs is just a part of the common changes in pain and depression comorbidities, such as chemical dysfunction and central nervous system changes.

It should be noted that although serotonin reuptake inhibitors (SSRIs) are first-line drugs for depression, but the efficacy of SSRIs in pain were disappointing. This difference raises a key question, whether to consider improving pain when treating depression in the context of pain. Previous studies have found that DA and NE play similar roles in depression, but NE plays a more important role in pain. On the one hand, this proves that SNRIs are better than SSRIs in the treatment of pain and depression comorbidities. On the other hand, in the treatment of depression in the context of pain, it is necessary to consider whether the drug can also improve the symptoms of pain.

3.2 Tricyclic Antidepressants (TCAs)

TCAs is a traditional antidepressant that acts by inhibiting the 5-HT transporter and NE transporter. They are similar to SNRIs in the treatment of depression. At the same time, there are evidences that TCAs can treat chronic pain, and interestingly, the types of pain treated by SNRIs and TCAs are very similar. A recent systematic review noted that the TCA amitriptyline and SNRIs duloxetine are first-line options for patients with fibromyalgia syndrome (FMS) (Hauser et al. 2012). Although SNRIs and TCAs are very similar in treating pain and depression, SNRIs are still more recommended because of the more serious side effect of TCAs (Attal et al. 2010).

3.3 *Ketamine*

Ketamine is a common dissociative anesthetic that has been used as an analgesic for decades. It is now considered a novel and promising drug for the treatment of comorbid pain and depression. In recent years, a large body of evidence indicates that a single low dose of ketamine has a rapid and effective therapeutic effect on depression and even treatment-resistant depression. The antidepressant effect can last for at least 24 h in most patients, and some patients can even last up to 7 days (Berman et al. 2000; Diazgranados et al. 2010; Price et al. 2009). Ketamine has long been used as an analgesic for acute pain and as an alternative therapy for chronic pain, and its efficacy in the latter case varies with the level of evidence (Hocking and Cousins 2003). The pain intensity of complex regional pain syndrome type I (CRPS I) was improved after the use of a ketamine-containing ointment (Ushida et al. 2002). The effectiveness of ketamine in the treatment of CRPS I was further confirmed by subsequent studies (Sigtermans et al. 2009). Although some side effects (such as hallucinations) have been observed during the treatment, overall it still outweighs the disadvantages. Importantly, preclinical studies on depression in the context of pain have shown that, in a rat model of CFA, both mechanical allodynia and depression-like behaviors were attenuated after a single dose of ketamine (20 mg/kg) injection (Zhang et al. 2016). It is plausible to assume that ketamine is a promising candidate drug that can effectively treat the pain and depression comorbidity. However, a litter different, in a rat model of SNI, a single subanesthetic dose of ketamine (10 mg/kg) did not improve the pain hypersensitivity in rats after SNI, but restored the depression-like behaviors of rats after SNI (Wang et al. 2011). This difference requires further experimentation to determine what mechanism ketamine works in both pain and depression.

Clinical studies also provide a promising reference. In one case report, a 14-year-old female patient suffered from severe depression and neuropathic pain, and multiple antidepressant treatments (fluoxetine, bupropion, and aripiprazole) and analgesic methods (ibuprofen, acetaminophen, lidocaine 5% patch, tizanidine, and oxycodone) failed. In detail, she had reported suicidal ideation and scored 7/10 on the numerical rating scale (NRS, 0–10 with 10 being the worst pain). However, 1 day after intravenous infusion of ketamine (7 μg/kg/min), depressive symptoms improved, with an NRS of 6/10 (no significant improvement). On day 5 after ketamine (4 μg/kg/min) intravenous infusion, NRS score was 0/10 (significant improvement) and the mood continued to improve. Symptom (pain and depression) relief persisted for 5 months after the patient's first intravenous ketamine. No obvious side effects (dysphoria or hallucinations) during the entire treatment (Weber et al. 2018).

In other case report, a 46-year-old man suffered from bipolar disorder (suicidal ideation) and chronic pain (shoulder), with quetiapine and bupropion prescriptions. On day 2 after ketamine (0.5 mg/kg) intramuscular injection, depressed mood was dramatically improved (PHQ-9 depression questionnaire score of 2) and pain severity score was 2–3/10 (Bigman et al. 2017). On the one hand, these two examples provide a reference for the future clinical use of ketamine to treat pain and depression comorbidity. On the other hand, it also shows that a reasonable dose of ketamine and real-time monitoring outweighs the risks in treatment-specific patients with pain and depression comorbidity.

A growing number of clinical and preclinical studies have shown that comorbid pain and depression are a complex problem with complicated and confused mechanisms. The imaging results suggest that there is a common brain region change in comorbidity. The animal model results suggest that chronic pain can eventually induce depression-like behavior. Further mechanism studies suggest that there are common neurotransmitter and neuromodulator changes in the process of comorbidity. As the research progresses, relevant therapeutic drugs continue to emerge, such as SNRIs and TCAs. Among them, ketamine has become a promising candidate for the treatment of comorbidities due to its eye-catching performance in both pain and depression. Ultimately, it may require a multidisciplinary approach and may actually need to consider and address the physical, psychological, and social factors that may affect the outcome from a global perspective.

References

Ackerman LL, Follett KA, Rosenquist RW (2003) Long-term outcomes during treatment of chronic pain with intrathecal clonidine or clonidine/opioid combinations. J Pain Symptom Manage 26(1):668–677

Aguera-Ortiz L, Failde I, Mico JA et al (2011) Pain as a symptom of depression: prevalence and clinical correlates in patients attending psychiatric clinics. J Affect Disord 130(1-2):106–112

Alba-Delgado C, Llorca-Torralba M, Horrillo I et al (2013) Chronic pain leads to concomitant noradrenergic impairment and mood disorders. Biol Psychiatry 73(1):54–62

Albert PR, Vahid-Ansari F, Luckhart C (2014) Serotonin-prefrontal cortical circuitry in anxiety and depression phenotypes: pivotal role of pre- and post-synaptic 5-HT1A receptor expression. Front Behav Neurosci 8:199

Almeida C, DeMaman A, Kusuda R et al (2015) Exercise therapy normalizes BDNF upregulation and glial hyperactivity in a mouse model of neuropathic pain. Pain 156(3):504–513

Apkarian AV, Bushnell MC, Treede RD et al (2005) Human brain mechanisms of pain perception and regulation in health and disease. Eur J Pain 9(4):463–484

Arnold LM (2007) Duloxetine and other antidepressants in the treatment of patients with fibromyalgia. Pain Med 8(Suppl 2):S63–S74

Attal N, Cruccu G, Baron R et al (2010) EFNS guidelines on the pharmacological treatment of neuropathic pain: 2010 revision. Eur J Neurol 17(9):1113–1e88

Bair MJ, Robinson RL, Katon W et al (2003) Depression and pain comorbidity: a literature review. Arch Intern Med 163(20):2433–2445

Baliki MN, Mansour A, Baria AT et al (2013) Parceling human accumbens into putative core and shell dissociates encoding of values for reward and pain. J Neurosci 33(41):16383–16393

Barbanti P, Fofi L, Aurilia C et al (2013) Dopaminergic symptoms in migraine. Neurol Sci 34(Suppl 1):S67–S70

Bazzichi L, Giannaccini G, Betti L et al (2006) Alteration of serotonin transporter density and activity in fibromyalgia. Arthritis Res Ther 8(4):R99

Berman RM, Cappiello A, Anand A et al (2000) Antidepressant effects of ketamine in depressed patients. Biol Psychiatry 47(4):351–354

Bigman D, Kunaparaju S, Bobrin B (2017) Use of ketamine for acute suicidal ideation in a patient with chronic pain on prescribed cannabinoids. BMJ Case Rep 2017

Blandini F, Rinaldi L, Tassorelli C et al (2006) Peripheral levels of BDNF and NGF in primary headaches. Cephalalgia 26(2):136–142

Brannan SK, Mallinckrodt CH, Brown EB et al (2005) Duloxetine 60 mg once-daily in the treatment of painful physical symptoms in patients with major depressive disorder. J Psychiatr Res 39(1):43–53

Bravo L, Torres-Sanchez S, Alba-Delgado C et al (2014) Pain exacerbates chronic mild stress-induced changes in noradrenergic transmission in rats. Eur Neuropsychopharmacol 24(6):996–1003

Bushnell MC, Ceko M, Low LA (2013) Cognitive and emotional control of pain and its disruption in chronic pain. Nat Rev Neurosci 14(7):502–511

Carr EC, Nicky Thomas V, Wilson-Barnet J (2005) Patient experiences of anxiety, depression and acute pain after surgery: a longitudinal perspective. Int J Nurs Stud 42(5):521–530

Chandran A, Iyo AH, Jernigan CS et al (2013) Reduced phosphorylation of the mTOR signaling pathway components in the amygdala of rats exposed to chronic stress. Prog Neuropsychopharmacol Biol Psychiatry 40:240–245

Cocksedge KA, Simon C, Shankar R (2014) A difficult combination: chronic physical illness, depression, and pain. Br J Gen Pract 64(626):440–441

Decosterd I, Woolf CJ (2000) Spared nerve injury: an animal model of persistent peripheral neuropathic pain. Pain 87(2):149–158

Delgado PL (2000) Depression: the case for a monoamine deficiency. J Clin Psychiatry 61(Suppl 6):7–11

DeVeaugh-Geiss AM, West SL, Miller WC et al (2010) The adverse effects of comorbid pain on depression outcomes in primary care patients: results from the ARTIST trial. Pain Med 11(5):732–741

Diazgranados N, Ibrahim L, Brutsche NE et al (2010) A randomized add-on trial of an N-methyl-D-aspartate antagonist in treatment-resistant bipolar depression. Arch Gen Psychiatry 67(8):793–802

Duric V, McCarson KE (2006a) Effects of analgesic or antidepressant drugs on pain- or stress-evoked hippocampal and spinal neurokinin-1 receptor and brain-derived neurotrophic factor gene expression in the rat. J Pharmacol Exp Ther 319(3):1235–1243

Duric V, McCarson KE (2006b) Persistent pain produces stress-like alterations in hippocampal neurogenesis and gene expression. J Pain 7(8):544–555

Eisenberger NI, Lieberman MD, Williams KD (2003) Does rejection hurt? An FMRI study of social exclusion. Science 302(5643):290–292

Fukuhara K, Ishikawa K, Yasuda S et al (2012) Intracerebroventricular 4-methylcatechol (4-MC) ameliorates chronic pain associated with depression-like behavior via induction of brain-derived neurotrophic factor (BDNF). Cell Mol Neurobiol 32(6):971–977

Gerrits MM, van Oppen P, van Marwijk HW, et al (2014) Pain and the onset of depressive and anxiety disorders. Pain 155(1): 53-59

Goffer Y, Xu D, Eberle SE et al (2013) Calcium-permeable AMPA receptors in the nucleus accumbens regulate depression-like behaviors in the chronic neuropathic pain state. J Neurosci 33(48):19034–19044

Hache G, Guiard BP, Nguyen TH et al (2015) Antinociceptive activity of the new triple reuptake inhibitor NS18283 in a mouse model of chemotherapy-induced neuropathic pain. Eur J Pain 19(3):322–333

Hauser W, Wolfe F, Tolle T et al (2012) The role of antidepressants in the management of fibromyalgia syndrome: a systematic review and meta-analysis. CNS Drugs 26(4):297–307

Hilderink PH, Burger H, Deeg DJ et al (2012) The temporal relation between pain and depression: results from the longitudinal aging study Amsterdam. Psychosom Med 74(9):945–951

Hocking G, Cousins MJ (2003) Ketamine in chronic pain management: an evidence-based review. Anesth Analg 97(6):1730–1739

Hooten WM (2016) Chronic pain and mental health disorders: shared neural mechanisms, epidemiology, and treatment. Mayo Clin Proc 91(7):955–970

Karege F, Perret G, Bondolfi G et al (2002) Decreased serum brain-derived neurotrophic factor levels in major depressed patients. Psychiatry Res 109(2):143–148

Kerr BJ, Bradbury EJ, Bennett DL et al (1999) Brain-derived neurotrophic factor modulates nociceptive sensory inputs and NMDA-evoked responses in the rat spinal cord. J Neurosci 19(12):5138–5148

Kim H, Chen L, Lim G et al (2012) Brain indoleamine 2,3-dioxygenase contributes to the comorbidity of pain and depression. J Clin Invest 122(8):2940–2954

Leino P, Magni G (1993) Depressive and distress symptoms as predictors of low back pain, neck-shoulder pain, and other musculoskeletal morbidity: a 10-year follow-up of metal industry employees. Pain 53(1):89–94

Li JX (2015) Pain and depression comorbidity: a preclinical perspective. Behav Brain Res 276:92–98

Li N, Lee B, Liu RJ et al (2010) mTOR-dependent synapse formation underlies the rapid antidepressant effects of NMDA antagonists. Science 329(5994):959–964

Li Q, Yue N, Liu SB et al (2014) Effects of chronic electroacupuncture on depression- and anxiety-like behaviors in rats with chronic neuropathic pain. Evid Based Complement Alternat Med 2014:158987

Mann JJ, Malone KM (1997) Cerebrospinal fluid amines and higher-lethality suicide attempts in depressed inpatients. Biol Psychiatry 41(2):162–171

Means-Christensen AJ, Roy-Byrne PP, Sherbourne CD et al (2008) Relationships among pain, anxiety, and depression in primary care. Depress Anxiety 25(7):593–600

Moriam S, Sobhani ME (2013) Epigenetic effect of chronic stress on dopamine signaling and depression. Genet Epigenet 5:11–16

Mutschler I, Ball T, Wankerl J et al (2012) Pain and emotion in the insular cortex: evidence for functional reorganization in major depression. Neurosci Lett 520(2):204–209

Ohayon MM (2004) Specific characteristics of the pain/depression association in the general population. J Clin Psychiatry 65(Suppl 12):5–9

Okubo M, Castro A, Guo W et al (2013) Transition to persistent orofacial pain after nerve injury involves supraspinal serotonin mechanisms. J Neurosci 33(12):5152–5161

Ordway GA (1997) Pathophysiology of the locus coeruleus in suicide. Ann N Y Acad Sci 836:233–252

Pearson-Fuhrhop KM, Dunn EC, Mortero S et al (2014) Dopamine genetic risk score predicts depressive symptoms in healthy adults and adults with depression. PLoS One 9(5):e93772

Piccinni A, Marazziti D, Catena M et al (2008) Plasma and serum brain-derived neurotrophic factor (BDNF) in depressed patients during 1 year of antidepressant treatments. J Affect Disord 105(1-3):279–283

Price RB, Nock MK, Charney DS et al (2009) Effects of intravenous ketamine on explicit and implicit measures of suicidality in treatment-resistant depression. Biol Psychiatry 66(5):522–526

Russo SJ, Nestler EJ (2013) The brain reward circuitry in mood disorders. Nat Rev Neurosci 14(9):609–625

Sagheddu C, Aroni S, De Felice M et al (2015) Enhanced serotonin and mesolimbic dopamine transmissions in a rat model of neuropathic pain. Neuropharmacology 97:383–393

Sanacora G, Treccani G, Popoli M (2012) Towards a glutamate hypothesis of depression: an emerging frontier of neuropsychopharmacology for mood disorders. Neuropharmacology 62(1):63–77

Sekine A, Hozumi S, Shimizu T (2016) Influence of painful physical symptoms in the treatment of Japanese patients with melancholic major depressive disorder: a prospective cohort study. Psychiatry Res 242:240–244

Sellmeijer J, Mathis V, Hugel S et al (2018) Hyperactivity of anterior cingulate cortex areas 24a/24b drives chronic pain-induced anxiodepressive-like consequences. J Neurosci 38(12):3102–3115

Shi M, Wang JY, Luo F (2010) Depression shows divergent effects on evoked and spontaneous pain behaviors in rats. J Pain 11(3):219–229

Sigtermans MJ, van Hilten JJ, Bauer MC et al (2009) Ketamine produces effective and long-term pain relief in patients with Complex Regional Pain Syndrome Type 1. Pain 145(3):304–311

Sobocki P, Jonsson B, Angst J et al (2006) Cost of depression in Europe. J Ment Health Policy Econ 9(2):87–98

Suzuki T, Amata M, Sakaue G et al (2007) Experimental neuropathy in mice is associated with delayed behavioral changes related to anxiety and depression. Anesth Analg 104(6):1570–1577. table of contents

Tateiwa H, Kawano T, Nishigaki A et al (2018) The role of hippocampal brain-derived neurotrophic factor in age-related differences in neuropathic pain behavior in rats. Life Sci 197:56–66

Tukey DS, Lee M, Xu D et al (2013) Differential effects of natural rewards and pain on vesicular glutamate transporter expression in the nucleus accumbens. Mol Brain 6:32

Ushida T, Tani T, Kanbara T et al (2002) Analgesic effects of ketamine ointment in patients with complex regional pain syndrome type 1. Reg Anesth Pain Med 27(5):524–528

Verne GN, Price DD, Callam CS et al (2012) Viscerosomatic facilitation in a subset of IBS patients, an effect mediated by N-methyl-D-aspartate receptors. J Pain 13(9):901–909

Von Korff M, Dworkin SF, Le Resche L et al (1988) An epidemiologic comparison of pain complaints. Pain 32(2):173–183

Wang J, Goffer Y, Xu D et al (2011) A single subanesthetic dose of ketamine relieves depression-like behaviors induced by neuropathic pain in rats. Anesthesiology 115(4):812–821

Weber G, Yao J, Binns S et al (2018) Case report of subanesthetic intravenous ketamine infusion for the treatment of neuropathic pain and depression with suicidal features in a pediatric patient. Case Rep Anesthesiol 2018:9375910

Wohlreich MM, Sullivan MD, Mallinckrodt CH et al (2009) Duloxetine for the treatment of recurrent major depressive disorder in elderly patients: treatment outcomes in patients with comorbid arthritis. Psychosomatics 50(4):402–412

Zhang GF, Wang J, Han JF et al (2016) Acute single dose of ketamine relieves mechanical allodynia and consequent depression-like behaviors in a rat model. Neurosci Lett 631:7–12

Zhao J, Seereeram A, Nassar MA et al (2006) Nociceptor-derived brain-derived neurotrophic factor regulates acute and inflammatory but not neuropathic pain. Mol Cell Neurosci 31(3):539–548

Distinct Roles of NMDA Receptor GluN2 Subunits in the Effects of Ketamine and Its Enantiomers

Soichiro Ide and Kazutaka Ikeda

Abstract The *N*-methyl-D-aspartate (NMDA) receptor antagonist ketamine is known to induce various pharmacological effects. It has attracted attention because of its rapid and sustained antidepressant effects in depressed patients, although its side effects have raised some concerns. NMDA receptors are considered to mediate various effects of ketamine, but other systems may be involved. NMDA receptors are di-heteromeric complexes that are composed of two GluN1 subunits and two homomeric GluN2A-D/3A subunits, or tri-heteromeric complexes that are composed of two GluN1 subunits and two heteromeric GluN2A-D/GluN3A, B subunits. The involvement of each NMDA receptor subunit in the neural mechanisms that underlie the various effects of ketamine remains unclear. This chapter focuses on GluN2 subunits and discusses the involvement of each GluN2 subunit in various pharmacological actions of ketamine.

Keywords GluN2 subunit · NMDA receptor · Ketamine · Phencyclidine (PCP) · NR2 subunit

1 Ketamine and NMDA Receptors

Ketamine and phencyclidine (PCP) have structural similarity and are classified as arylcycloalkylamines. They act on *N*-methyl-D-aspartate (NMDA) receptors as non-competitive antagonists. Ketamine has been widely used in humans and animals and is listed as an essential medicine by the World Health Organization (WHO) as an anesthetic agent (World Health Organization 2017). The recreational use of ketamine (street names: "K," "Special K," etc.) has spread in many parts of the world,

S. Ide · K. Ikeda (✉)
Addictive Substance Project, Tokyo Metropolitan Institute of Medical Science, Setagaya-ku, Tokyo, Japan
e-mail: ikeda-kz@igakuken.or.jp

K. Hashimoto et al. (eds.), *Ketamine*,
https://doi.org/10.1007/978-981-15-2902-3_10

and new problems have emerged, including physical harm and addiction (Morgan and Curran 2012). Treatment with either ketamine or PCP can mimic symptoms of schizophrenia, thus promoting genetic, biochemical, and pharmacological research on the NMDA receptor hypofunction theory of schizophrenia (Coyle et al. 2003; Kantrowitz and Javitt 2010). NMDA receptors have been suggested to be required for recognition memory (de Lima et al. 2005; Rampon et al. 2000; Winters and Bussey 2005). High-dose ketamine or PCP administration is well known to cause cognitive impairment. Furthermore, recent studies have revealed that subanesthetic doses of ketamine exert rapid and sustained antidepressant effects (Berman et al. 2000; Garcia et al. 2008; Krystal et al. 2013; Price et al. 2009) and relieve suicidal ideation (Mallick and McCullumsmith 2016) and symptoms of posttraumatic disorders (Feder et al. 2014).

Major depressive disorder is a serious mental disorder that affects approximately 16% of the population worldwide, causing serious health consequences (Kessler et al. 2003). Although such interventions as pharmacotherapy and cognitive behavioral psychotherapy are available, most patients remain resistant to treatment. Furthermore, existing monoamine agonist-based drug therapies often take weeks or months to exert their full therapeutic effects (Insel and Wang 2009). Thus, antidepressants with novel mechanisms are being sought. Ketamine is considered to exert rapid and sustained antidepressant effects, thus attracting much research attention. Although many studies have proposed several mechanisms of action, still unclear is the way in which ketamine produces its various effects.

1.1 Mechanisms of Action of Ketamine

NMDA receptors are thought to mainly mediate the effects of ketamine (Zanos and Gould 2018), although recent studies suggest the potential relevance of other neurotransmitter systems, including dopamine (Belujon and Grace 2014; Can et al. 2016; Kapur and Seeman 2002), serotonin (Kapur and Seeman 2002), opioid (Williams et al. 2018; Gupta et al. 2011), sigma (Robson et al. 2012), γ-aminobutyric acid (GABA; McNally et al. 2011), and α-amino-3-hydroxy-5-methyl-4-isoxazolepropionic acid (AMPA; Zanos et al. 2016) receptors. Ketamine directly binds to NMDA receptors as a noncompetitive antagonist. Thus, it may be expected to block excitatory glutamatergic neurotransmission via NMDA receptor inhibition. A recent study found that ketamine decreased the NMDA receptor-dependent burst firing of glutamatergic neurons in the lateral habenula (LHb) that were transiently activated by aversive stimuli (Hikosaka 2010). This ketamine-induced reduction of bursting activity in the LHb was associated with an acute antidepressant effect in a rat model of depression (Yang et al. 2018). With regard to the acute effects of ketamine, particularly its antidepressant effects, the blockade of NMDA receptors that are expressed on LHb neurons may play an important role. Subanesthetic doses of ketamine were reported to increase the overall activity in the prefrontal cortex in healthy volunteers (Breier et al. 1997), which was hypothesized to be attributable to the inhibition of NMDA receptors that are expressed on GABAergic interneurons (Moghaddam et al. 1997; Homayoun and Moghaddam 2007). Subanesthetic doses of ketamine also significantly increased extracellular glutamate levels in rats (Moghaddam et al. 1997). Based on the higher frequency of interneuron firing compared with pyramidal neurons (Neske et al. 2015), which increases relief of Mg^{2+} blockade in NMDA receptors, ketamine may selectively bind at the channel pore of NMDA receptors on interneurons (Seamans 2008). The inhibition of NMDA receptors on GABAergic interneurons and concomitant pyramidal cell disinhibition and the enhancement of excitatory glutamatergic neurotransmission in the medial prefrontal cortex and other brain regions may be involved in the mechanism of action of various effects of ketamine.

Ketamine is also considered to exert its effects through various NMDA receptor mechanisms. One such mechanism is the inhibition of spontaneous resting-state neurotransmission via NMDA receptors. Ketamine was reported to block miniature excitatory postsynaptic currents (mEPSCs), which tonically suppresses protein synthesis (Sutton et al. 2007) and was reported to induce the disinhibition of protein synthesis and lead to synaptic potentiation in the hippocampus and subsequently antidepressant effects (Autry et al. 2011; Nosyreva et al. 2013). Ketamine has also been reported to inhibit NMDA receptor mEPSCs under the conditions of physiological levels of Mg^{2+} (Gideons et al. 2014). Thus, the effects of ketamine are hypothesized to be expressed partially through the inhibition of spontaneous NMDA receptor-mediated neurotransmission and the concomitant enhancement of synaptic neurotransmission through a protein synthesis-dependent mechanism. The direct inhibition of extrasynaptic NMDA receptors is another possible mechanism of action of ketamine. The existence of extrasynaptic NMDA receptors that are not located in the postsynaptic density has been reported in both immunohistochemical and electrophysiological studies (Hardingham and Bading 2010). Extrasynaptic NMDA receptors are not activated by typical transient synaptic glutamate release; instead, they are chronically activated by low levels of ambient glutamate within the extracellular space (Rothstein et al. 1996). Ketamine is hypothesized to specifically inhibit extrasynaptic NMDA receptors, thus preventing ambient glutamate-induced tonic activation of these receptors and the concomitant induction of pyramidal neuron excitation that disinhibits protein synthesis (Miller et al. 2014; Zanos and Gould 2018).

1.2 NMDA Receptor Subunits

NMDA receptors are di-heteromeric complexes that are composed of two GluN1 subunits and two homomeric GluN2A-D/3A subunits, or tri-heteromeric complexes that are composed of two GluN1 subunits and two heteromeric GluN2A-D/GluN3A, B subunits (Fig. 1a; Paoletti et al. 2013). These subunits are critical determinants of NMDA receptor heterogeneity and provide the major influence in determining the biophysical, pharmacological, and signaling properties of different NMDA receptor subtypes (Paoletti et al. 2013; Wyllie et al. 2013). Each NMDA receptor subunit is composed of four discrete semiautonomous domains (Traynelis et al. 2010), including the extracellular amino-terminal domain (ATD), extracellular ligand-binding domain (LBD), transmembrane domain (TMD), and intracellular carboxyl-terminal domain (CTD). Ketamine and PCP are thought to bind to the TMD, which represents the most highly conserved portion of NMDA receptor subunits (Fig. 1b). Further details about the site of action of ketamine on NMDA receptors are provided in another chapter in this book. An in vitro study that used oocytes that expressed NMDA receptor subunits found that the potencies of compounds that are classified as arylcycloalkylamines varied between four GluN2 subunits (Dravid et al. 2007). Ketamine was reported to be more potent at the GluN1/GluN2B subunit than at the GluN1/GluN2A, GluN1/GluN2C, and GluN1/GluN2D subunits, whereas PCP had less potency at the GluN1/GluN2A subunit than at the GluN1/GluN2B and GluN1/GluN2C subunits. (+)-MK-801 was almost equipotent across all four di-heteromeric GluN2 subunit-containing NMDA receptors. Furthermore, the GluN2 ATD was reported to regulate agonist potency, the deactivation time course, the channel open probability, and the mean channel open/closed duration of different GluN2 subunits (Gielen et al. 2009; Yuan et al. 2009). Thus, differences in subunit compositions of NMDA receptors likely explain the diversity of actions of various noncompetitive NMDA receptor antagonists, including ketamine.

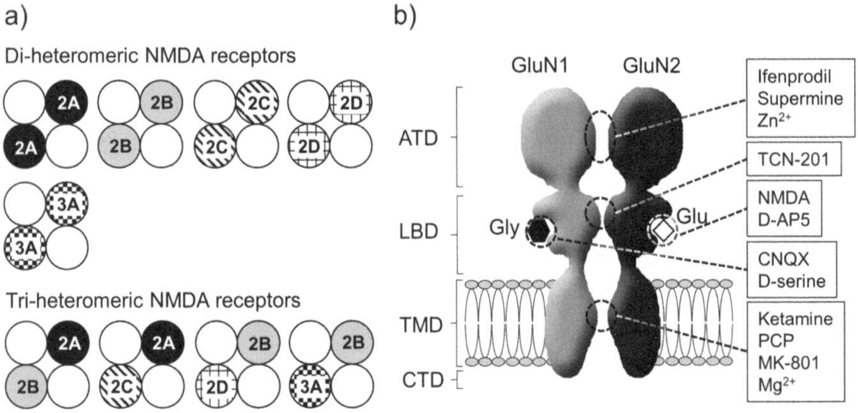

Fig. 1 Diversity and structure of NMDA receptors. (**a**) Schematic representation of di-heteromeric and tri-heteromeric NMDA receptors with GluN1 subunits (indicated in white circle) that are thought to exist in the central nervous system. (**b**) Schematic of the general structure of NMDA receptors and multiple binding sites for extracellular ligands. *ATD* extracellularly located amino terminal domain, *LBD* ligand-binding domain, *TMD* transmembrane domain, *CTD* intracellularly located C-terminal domain

1.3 Enantiomers of Ketamine

Ketamine is a racemic mixture of equal amounts of the enantiomers *R*(-)-ketamine and *S*(+)-ketamine (hereinafter denoted by [*R*]-ketamine and [*S*]-ketamine, respectively). (*S*)-ketamine has been thought to be an active isomer because of its higher affinity for NMDA receptors and greater anesthetic potency (Domino 2010; Franks and Lieb 1994; Murray and Leid 1984). (*S*)-ketamine was also reported to be more potent than (*R*)-ketamine as an analgesic (Klepstad et al. 1990) and in a drug discrimination study (Brady and Balster 1982). A recent clinical trial reported the rapid onset of robust antidepressant effects of (*S*)-ketamine in patients with treatment-resistant depression (Singh et al. 2016). However, (*R*)-ketamine has also been reported to exert more potent and sustained antidepressant effects than (*S*)-ketamine, without causing such adverse effects as psychotomimetic behaviors, neurotoxicity, and abuse potential, in animal models (Fukumoto et al. 2017; Yang et al. 2015; Zhang et al. 2014b). Details of differences in both enantiomers of ketamine are described in another chapter in this book. The neural mechanisms and role of each GluN2 subunit that underlie these differential effects of ketamine and its enantiomers remain unclear.

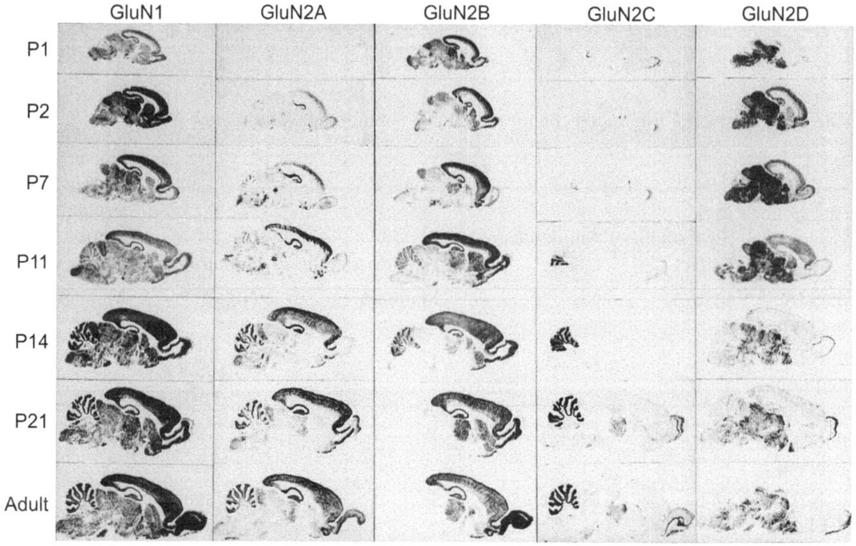

Fig. 2 Expression of NMDA receptor subunits. Autoradiograms obtained by in situ hybridization of oligonucleotide probes to parasagittal sections of rat brain on indicated postnatal (P) days. (The figure is modified from Akazawa et al. (1994) with permission from *Journal of Comparative Neurology*)

2 NMDA Receptor GluN2 Subunits

GluN2 subunits exhibit a developmentally and spatially regulated expression pattern (Fig. 2; Akazawa et al. 1994; Monyer et al. 1994; Watanabe et al. 1992). In the embryonic rodent brain, GluN2B and GluN2D expression is prominent. The latter is mostly found in caudal regions. Strong GluN2A expression begins around the second postnatal week of development. GluN2A and GluN2B subunits are the predominantly expressed NMDA receptor subunits in the adult forebrain. Together with this progressive increase in GluN2A expression, GluN2D expression markedly decreases postnatally and is expressed at low levels in the adult brain, mostly in the diencephalon and mesencephalon, although functional GluN2D-containing NMDA receptors are present at older ages, particularly in the basal ganglia (Wyllie et al. 2013). GluN2C expression appears late in development (postnatal day 10), and its expression is mainly confined to the cerebellum, thalamus, and olfactory bulb. The specific expression of GluN2B and GluN2D subunits early in development suggests that these subunits are important for synaptogenesis and synaptic maturation. In the adult central nervous system, particularly in higher brain structures (e.g., hippocampus and cortex), GluN2A and GluN2B are the predominant subunits (Akazawa et al. 1994; Monyer et al. 1994; Watanabe et al. 1992), indicating that they could play central roles in synaptic function and plasticity. The specific roles of each NMDA receptor subunit in various effects of ketamine remain unclear. Gene knockout studies of different NMDA receptor subunits have suggested that each plays a different role in the pharmacological effects of ketamine (Table 1). Below we discuss the involvement of each GluN2 subunit in various pharmacological actions of ketamine.

Table 1 Ketamine/PCP-induced changes in GluN2 subunit knockout mice

	Anesthetic effect	Antidepressant effect	Hyperlocomotion	Cognitive impairment	Dopamine release in striatum/prefrontal cortex
GluN2A knockout	↓	NR	Maintain	NR	Maintain
GluN2B conditional knockout	NR	↓	↓	↓	NR
GluN2C knockout	NR	NR	↑	↑	NR
GluN2D knockout	NR	Maintain[a]	↓	Maintain[a]	↓

The data are described in a simple form with reference to the following papers: Sato et al. (2005), Brigman et al. (2013), Miller et al. (2014), Gupta et al. (2016), Hagino et al. (2010), Yamamoto et al. (2013, 2016), Sapkota et al. (2016), and Ide et al. (2017, 2019). *NR* not reported

[a]Sustained antidepressant effect and cognitive impairment induced by (*R*)-ketamine were lower (Ide et al. 2017, 2019)

2.1 Ketamine and the GluN2A Subunit

NMDA receptors that contain the GluN2A subunit (also known as ε1 or NR2A) have the fastest deactivation time course, higher single-channel open probability, and higher Mg^{2+} sensitivity compared with other GluN2 subunits (Traynelis et al. 2010; Wyllie et al. 2013). Previous studies have investigated the role of GluN2A subunits in the anesthetic effect of ketamine. GluN2A knockout mice were resistant to the hypnotic effect of intraperitoneal ketamine administration, reflected by a decrease in the loss of righting reflex (Sato et al. 2005). GluN2A knockout mice were also reported to be resistant to the hypnotic effect of pentobarbital (Petrenko et al. 2004), which is a barbituric acid derivative that acts via the potentiation of $GABA_A$ receptors. The dysfunction of GABAergic neurotransmission was also speculated in GluN2A knockout mice because NMDA-stimulated GABA release was markedly lower in striatal slices from GluN2A knockout mice (Miyamoto et al. 2001). Thus, the decrease in the hypnotic effect of ketamine in GluN2A knockout mice is considered to be partially attributable to compensatory changes; thus, the specific contribution of GluN2A subunit-containing NMDA receptors to the anesthetic effect of ketamine is not completely understood (Petrenko et al. 2014).

Although GluN2A-containing NMDA receptors are the major NMDA receptor subtype in the adult brain (Paoletti et al. 2013), the role of GluN2A subunits in the actions of subanesthetic doses of ketamine have rarely been tested. A recent study found that the rapid enhancement of visual cortical responses and gamma-band oscillations by subanesthetic doses of ketamine were abolished in GluN2A knockout mice (Picard et al. 2019). This study also found that parvalbumin-expressing interneuron-selective conditional knockout mice exhibited abolishment of the rapid enhancement of visual cortical responses and gamma-band oscillations by ketamine. Although the possibility that compensatory changes occurred in GluN2A knockout mice cannot necessarily be discarded, GluN2A-containing NMDA receptors, especially on parvalbumin-expressing interneurons, could be involved in some of the acute effects of ketamine. Moreover, the expression of both GluN2A and GluN2B subunits was reported to be lower in the prefrontal cortex in postmortem brain samples in major depression patients compared with healthy controls (Feyissa et al. 2009). GluN2A knockout mice were also reported to present a reduction of depressive-related behaviors in the forced swim test and tail suspension test (Boyce-Rustay and Holmes 2006). Thus, the GluN2A subunit may be involved in the onset or pathophysiology of depression. Further elucidation of the role of GluN2A subunits in the antidepressant effects of ketamine is needed.

2.2 Ketamine and the GluN2B Subunit

NMDA receptors that contain the GluN2B (ε2 or NR2B) subunit have a faster deactivation time course, higher single-channel open probability, and higher Mg^{2+} sensitivity compared with the other GluN2 subunits (Traynelis et al. 2010; Wyllie et al.

2013). The GluN2B subunit, together with the GluN2A subunit, is the most abundant subunit in the adult brain. The GluN2A and GluN2B subunits play different and important roles in synaptic activity in the central nervous system (Bartlett et al. 2007; Kocsis 2012; Anastasio et al. 2009). The systemic administration of selective antagonists of GluN2B-containing NMDA receptors, such as CP-101,606, ifenprodil, and Ro 25-6981, produced antidepressant-like effects in rodents (Li et al. 2010, 2011; Lima-Ojeda et al. 2013; Maeng et al. 2008) and had therapeutic efficacy in human patients with treatment-refractory major depressive disorder (Preskorn et al. 2008), although the effects of these antagonists were not as sustained as ketamine. With regard to the relationship between the GluN2B subunit and depression, gene linkage analysis in humans confirmed a role specifically for the GluN2B subunit in bipolar disorder (Martucci et al. 2006). A recent study also reported significant differences in allele and genotype frequencies between treatment-resistant depression and control groups for the rs1805502 polymorphism within the *GRIN2B* gene that encodes the GluN2B subunit (Zhang et al. 2014a). Thus, GluN2B subunits may play a pivotal role in the antidepressant effect of ketamine. Furthermore, Ro 25-6981 but not the GluN2A-preferring antagonist NVP-AAM077 substituted for PCP in a drug discrimination test, disrupted prepulse inhibition, and induced hyperlocomotion in rats (Chaperon et al. 2003). Ro 25-6981 and PCP also substituted for ketamine in drug discrimination tests in rats (De Vry and Jentzsch 2003). Although some differences are likely, the pharmacological actions of GluN2B antagonists might be similar to ketamine. Thus, studies of GluN2B-containing NMDA receptor-selective antagonists have suggested that this receptor subtype could be involved in the various acute effects of ketamine.

Unlike the GluN2A subunit, genetic deletion of the GluN2B subunit is fatal by postnatal day 0 (Kutsuwada et al. 1996). Recent studies of region-specific GluN2B conditional knockout mice have reported novel findings. Corticostriatal or striatal GluN2B deletion and Ro 25-6981-induced GluN2B antagonism in the dorsal striatum impaired choice learning in mice, whereas cortical GluN2B deletion and GluN2B antagonism in the orbitofrontal cortex impaired shifting performance (Brigman et al. 2013). Together with the results of previous studies that evaluated selective GluN2B antagonists (Dix et al. 2010; Higgins et al. 2005), the GluN2B subunit is suggested to be crucial for mediating certain types of cognitive functions and may play a pivotal role in cognitive impairment that is induced by ketamine and PCP. Furthermore, cortex- and principal neuron-specific GluN2B knockout mice have been reported to recapitulate the actions of ketamine on depressive-like behavior, excitatory synaptic transmission, mechanistic/mammalian target of rapamycin (mTOR) activation, and synaptic protein synthesis. These conditional knockout mice were also reported to exhibit the loss of these effects of ketamine (Miller et al. 2014). These data suggest a role for GluN2B-containing NMDA receptors in the rapid antidepressant actions of ketamine, based on their ability to directly suppress mTOR signaling and limit protein synthesis in principal cortical neurons.

2.3 Ketamine and the GluN2C Subunit

NMDA receptors that contain the GluN2C (ε3 or NR2C) subunit have a slower deactivation time course, lower single-channel open probability, and lower Mg^{2+} sensitivity compared with GluN2A and GluN2B subunits (Traynelis et al. 2010; Wyllie et al. 2013). Although the expression of GluN2C subunits is considerably more restricted than the GluN2A and GluN2B subunits, ketamine at an estimated concentration that has psychotogenic actions in humans was speculated to block a substantial proportion of GluN2C-containing NMDA receptors but has less effect on GluN2A and GluN2B subunits (Khlestova et al. 2016). Dravid et al. (2007) suggested that the behavioral effects of ketamine are more likely to be mediated by GluN2C- or GluN2D-containing NMDA receptors than by GluN2A- or GluN2B-containing NMDA receptors, based on an in vitro study of NMDA receptor subunit expression in oocytes. Ketamine treatment in neonatal rats on postnatal day 7 increased the mRNA and protein expression of the GluN2C subunit, although these changes were also shown for other subunits (Han et al. 2010; Shi et al. 2010). GluN2C knockout mice were reported to present no difference in spontaneous activity, basal anxiety, immobility in the forced swim test, novel object recognition, pain sensitivity, or reference memory compared with wild-type mice, but these knockout mice exhibited deficits in fear acquisition and working memory (Hillman et al. 2011). Additionally, hyperresponsiveness to PCP-induced social and cognitive deficits, alterations of basal neuronal oscillations, and the MK-801-induced augmentation of neuronal oscillations were observed in GluN2C knockout mice, and these findings may be relevant to the pathophysiology of schizophrenia (Gupta et al. 2016). Lower GluN2C subunit expression was also observed in the prefrontal cortex and thalamus in postmortem brains from patients with schizophrenia but not from patients with major depression (Akbarian et al. 1996; Beneyto and Meador-Woodruff 2008). Thus, GluN2C subunit-containing NMDA receptors could be thought to play a specific role in producing psychotogenic effects of high-dose ketamine.

2.4 Ketamine and the GluN2D Subunit

NMDA receptors that contain the GluN2D (ε4 or NR2D) subunit have the slowest deactivation time course, lower single-channel open probability, and lower Mg^{2+} sensitivity compared with GluN2A and GluN2B subunits (Traynelis et al. 2010; Wyllie et al. 2013). The expression of GluN2D subunits is highest during early development, with a low-level and restricted expression in the adult brain. No selective agonists or antagonists of GluN2D subunits are yet available, thus hindering research on the role of GluN2D subunits in the effects of ketamine. Recent studies reported the expression of functional GluN2D-containing NMDA receptors at older ages (Swanger et al. 2015; Tozzi et al. 2016; Zhang and Chergui 2015). The selective GluN2C/D potentiator CIQ and antagonist DQP-1105 were shown to modulate neurotransmission in basal ganglia regions in adult rodents (Swanger et al. 2015; Zhang et al. 2014c), but no evidence of GluN2C expression has been reported in these brain regions. GluN2D knockout mice exhibited impairments in spatial memory acquisition, an increase in depressive-like behavior in the tail suspension test, lower sucrose preference, and a reduction of parvalbumin-immunopositive neurons compared with wild-type mice (Sapkota et al. 2016; Yamamoto et al. 2017). We previously reported that PCP significantly increased locomotor activity, caused motor impairment, and increased extracellular dopamine levels in wild-type mice but not in GluN2D knockout mice (Hagino et al. 2010; Yamamoto et al. 2013). Our group and others also reported that GluN2D knockout mice did not develop ketamine-induced hyperlocomotion or locomotor sensitization (Sapkota et al. 2016; Yamamoto et al. 2016). Sapkota et al. (2016) reported attenuation of the ketamine-induced increase in cortical gamma-band oscillatory power in GluN2D knockout mice compared with wild-type mice. Our group also found that the antidepressant effects of ketamine were preserved in GluN2D knockout mice that exhibited restraint stress-induced depressive-like behavior (Ide et al. 2017). These results suggest that GluN2D-containing NMDA receptors play a critical role in neuronal oscillation and psychotomimetic effects of ketamine but do not play a role in antidepressant effects of ketamine.

Unlike the other GluN2 subunits, differential roles of the GluN2D subunit in the effects of ketamine enantiomers have been reported. We recently found that the GluN2D subunit plays an important role in the sustained but not rapid antidepressant effects of (*R*)-ketamine, whereas this subunit does not appear to be involved in the antidepressant effects of (*RS*)-ketamine or (*S*)-ketamine (Ide and Ikeda 2018; Ide et al. 2017). A pivotal role for the GluN2D subunit in cognitive impairment that is induced by (*R*)-ketamine but not (*RS*)-ketamine or (*S*)-ketamine was also suggested (Ide et al. 2019). Yang et al. (2016) recently showed that the repeated intermittent administration of (*S*)-ketamine (10 mg/kg once weekly for 8 weeks) but not (*R*)-ketamine caused the loss of parvalbumin immunoreactivity in the medial prefrontal cortex and hippocampus in mice (Yang et al. 2016). Thus, chronic or repeated ketamine treatment appears to reduce the number of parvalbumin-positive GABAergic interneurons in the prefrontal cortex and hippocampus, resulting in cognitive impairment. Although expression of the GluN2D subunit has been found in presynaptic axons of many GABAergic neurons, GluN2D mRNA expression is mainly restricted to diencephalic, mesencephalic, and brainstem structures (Watanabe et al. 1992). We previously showed that PCP increased the number of Fos-positive neurons in the striatum and prefrontal cortex in wild-type mice but not in GluN2D knockout mice (Yamamoto et al. 2013). Therefore, PCP and ketamine may inhibit presynaptic GluN2D-containing NMDA receptors on GABAergic neurons that project to the prefrontal cortex, thus decreasing GABA release and activating the prefrontal cortex–dorsal striatum pathway. This activation may underlie the acute psychiatric effects of PCP and (*R*)-ketamine. Altogether, ketamine and its enantiomers appear to induce cognitive impairment, locomotor sensitization, and rapid and sustained antidepressant effects through different mechanisms, including a mechanism that involves NMDA receptors that contain the GluN2D subunit.

Future studies should investigate the specific roles of each GluN2 subunit in the effects of ketamine and its enantiomers.

3 Conclusion

Ketamine is well known to elicit various pharmacological responses, but the role of each GluN2 subunit in these effects is still not well known. Studies that used subunit-selective NMDA receptor modulators and gene knockout animals suggest that NMDA receptors that contain GluN2B subunits are involved in ketamine-induced antidepressant effects, and NMDA receptors that contain GluN2B/D subunits are involved in ketamine-induced psychotomimetic effects, but possible contributions of other subunits need to be considered. Although ketamine can provide rapid and sustained antidepressant responses, unclear is why other inhibitors of NMDA receptors do not produce similar effects. In humans and animals, different behavioral effects are triggered by different doses of ketamine, with different time courses. Depending on dose, time, and the depolarization state of neurons, different targets likely participate in different behavioral effects. The differential affinity of NMDA receptor ligands for NMDA receptors that contain different subunits, combined with different pharmacodynamics and pharmacokinetics, results in various pharmacological effects, with the potential involvement of other neuronal systems. Progress in the development of selective positive and negative allosteric modulators of each di-heteromeric and tri-heteromeric NMDA receptor will further elucidate the mechanism of action of various pharmacological effects of ketamine and its enantiomers and improve the NMDA receptor-targeted treatment of depression and various other neuropathological and psychiatric conditions.

Acknowledgment The authors thank Michael Arends for English editing.

Conflict of Interest: Authors declare no potential conflicts of interest.

Funding: This work was supported by JSPS KAKENHI (16K15565, 16H06276, and 17H04324 to K. I. and 17K08612 to S. I.), MEXT KAKENHI (16H06276 [AdAMS] to K. I.), AMED (JP19dk0307071, JP18mk0101076, and JP19ek0610011 to K. I.), and the Smoking Research Foundation (K. I.).

References

Akazawa C, Shigemoto R, Bessho Y, Nakanishi S, Mizuno N (1994) Differential expression of five N-methyl-D-aspartate receptor subunit mRNAs in the cerebellum of developing and adult rats. J Comp Neurol 347(1):150–160. https://doi.org/10.1002/cne.903470112

Akbarian S, Sucher NJ, Bradley D, Tafazzoli A, Trinh D, Hetrick WP, Potkin SG, Sandman CA, Bunney WE Jr, Jones EG (1996) Selective alterations in gene expression for NMDA receptor subunits in prefrontal cortex of schizophrenics. J Neurosci 16(1):19–30

Anastasio NC, Xia Y, O'Connor ZR, Johnson KM (2009) Differential role of N-methyl-D-aspartate receptor subunits 2A and 2B in mediating phencyclidine-induced perinatal neuronal apoptosis and behavioral deficits. Neuroscience 163(4):1181–1191. https://doi.org/10.1016/j.neuroscience.2009.07.058

Autry AE, Adachi M, Nosyreva E, Na ES, Los MF, Cheng PF, Kavalali ET, Monteggia LM (2011) NMDA receptor blockade at rest triggers rapid behavioural antidepressant responses. Nature 475(7354):91–95. https://doi.org/10.1038/nature10130

Bartlett TE, Bannister NJ, Collett VJ, Dargan SL, Massey PV, Bortolotto ZA, Fitzjohn SM, Bashir ZI, Collingridge GL, Lodge D (2007) Differential roles of NR2A and NR2B-containing NMDA receptors in LTP and LTD in the CA1 region of two-week old rat hippocampus. Neuropharmacology 52(1):60–70. https://doi.org/10.1016/j.neuropharm.2006.07.013

Belujon P, Grace AA (2014) Restoring mood balance in depression: ketamine reverses deficit in dopamine-dependent synaptic plasticity. Biol Psychiatry 76(12):927–936. https://doi.org/10.1016/j.biopsych.2014.04.014

Beneyto M, Meador-Woodruff JH (2008) Lamina-specific abnormalities of NMDA receptor-associated postsynaptic protein transcripts in the prefrontal cortex in schizophrenia and bipolar disorder. Neuropsychopharmacology 33(9):2175–2186. https://doi.org/10.1038/sj.npp.1301604

Berman RM, Cappiello A, Anand A, Oren DA, Heninger GR, Charney DS, Krystal JH (2000) Antidepressant effects of ketamine in depressed patients. Biol Psychiatry 47(4):351–354

Boyce-Rustay JM, Holmes A (2006) Genetic inactivation of the NMDA receptor NR2A subunit has anxiolytic- and antidepressant-like effects in mice. Neuropsychopharmacology 31(11):2405–2414. https://doi.org/10.1038/sj.npp.1301039

Brady KT, Balster RL (1982) Discriminative stimulus properties of ketamine stereoisomers in phencyclidine-trained rats. Pharmacol Biochem Behav 17(2):291–295

Breier A, Malhotra AK, Pinals DA, Weisenfeld NI, Pickar D (1997) Association of ketamine-induced psychosis with focal activation of the prefrontal cortex in healthy volunteers. Am J Psychiatry 154(6):805–811. https://doi.org/10.1176/ajp.154.6.805

Brigman JL, Daut RA, Wright T, Gunduz-Cinar O, Graybeal C, Davis MI, Jiang Z, Saksida LM, Jinde S, Pease M, Bussey TJ, Lovinger DM, Nakazawa K, Holmes A (2013) GluN2B in corticostriatal circuits governs choice learning and choice shifting. Nat Neurosci 16(8):1101–1110. https://doi.org/10.1038/nn.3457

Can A, Zanos P, Moaddel R, Kang HJ, Dossou KS, Wainer IW, Cheer JF, Frost DO, Huang XP, Gould TD (2016) Effects of ketamine and ketamine metabolites on evoked striatal dopamine release, dopamine receptors, and monoamine transporters. J Pharmacol Exp Ther 359(1):159–170. https://doi.org/10.1124/jpet.116.235838

Chaperon F, Muller W, Auberson YP, Tricklebank MD, Neijt HC (2003) Substitution for PCP, disruption of prepulse inhibition and hyperactivity induced by N-methyl-D-aspartate receptor antagonists: preferential involvement of the NR2B rather than NR2A subunit. Behav Pharmacol 14(5-6):477–487. https://doi.org/10.1097/01.fbp.0000091471.79060.ed

Coyle JT, Tsai G, Goff D (2003) Converging evidence of NMDA receptor hypofunction in the pathophysiology of schizophrenia. Ann N Y Acad Sci 1003:318–327

de Lima MN, Laranja DC, Bromberg E, Roesler R, Schroder N (2005) Pre- or post-training administration of the NMDA receptor blocker MK-801 impairs object recognition memory in rats. Behav Brain Res 156(1):139–143. https://doi.org/10.1016/j.bbr.2004.05.016

De Vry J, Jentzsch KR (2003) Role of the NMDA receptor NR2B subunit in the discriminative stimulus effects of ketamine. Behav Pharmacol 14(3):229–235. https://doi.org/10.1097/01.fbp.0000069577.37661.ce

Dix S, Gilmour G, Potts S, Smith JW, Tricklebank M (2010) A within-subject cognitive battery in the rat: differential effects of NMDA receptor antagonists. Psychopharmacology (Berl) 212(2):227–242. https://doi.org/10.1007/s00213-010-1945-1

Domino EF (2010) Taming the ketamine tiger 1965. Anesthesiology 113(3):678–684. https://doi.org/10.1097/ALN.0b013e3181ed09a2

Dravid SM, Erreger K, Yuan H, Nicholson K, Le P, Lyuboslavsky P, Almonte A, Murray E, Mosely C, Barber J, French A, Balster R, Murray TF, Traynelis SF (2007) Subunit-specific mechanisms

and proton sensitivity of NMDA receptor channel block. J Physiol 581(Pt 1):107–128. https://doi.org/10.1113/jphysiol.2006.124958
Feder A, Parides MK, Murrough JW, Perez AM, Morgan JE, Saxena S, Kirkwood K, Aan Het Rot M, Lapidus KA, Wan LB, Iosifescu D, Charney DS (2014) Efficacy of intravenous ketamine for treatment of chronic posttraumatic stress disorder: a randomized clinical trial. JAMA Psychiat 71(6):681–688. https://doi.org/10.1001/jamapsychiatry.2014.62
Feyissa AM, Chandran A, Stockmeier CA, Karolewicz B (2009) Reduced levels of NR2A and NR2B subunits of NMDA receptor and PSD-95 in the prefrontal cortex in major depression. Prog Neuropsychopharmacol Biol Psychiatry 33(1):70–75. https://doi.org/10.1016/j.pnpbp.2008.10.005
Franks NP, Lieb WR (1994) Molecular and cellular mechanisms of general anaesthesia. Nature 367(6464):607–614. https://doi.org/10.1038/367607a0
Fukumoto K, Toki H, Iijima M, Hashihayata T, Yamaguchi JI, Hashimoto K, Chaki S (2017) Antidepressant potential of (R)-ketamine in rodent models: comparison with (S)-ketamine. J Pharmacol Exp Ther 361(1):9–16. https://doi.org/10.1124/jpet.116.239228
Garcia LS, Comim CM, Valvassori SS, Reus GZ, Barbosa LM, Andreazza AC, Stertz L, Fries GR, Gavioli EC, Kapczinski F, Quevedo J (2008) Acute administration of ketamine induces antidepressant-like effects in the forced swimming test and increases BDNF levels in the rat hippocampus. Prog Neuropsychopharmacol Biol Psychiatry 32(1):140–144. https://doi.org/10.1016/j.pnpbp.2007.07.027
Gideons ES, Kavalali ET, Monteggia LM (2014) Mechanisms underlying differential effectiveness of memantine and ketamine in rapid antidepressant responses. Proc Natl Acad Sci U S A 111(23):8649–8654. https://doi.org/10.1073/pnas.1323920111
Gielen M, Siegler Retchless B, Mony L, Johnson JW, Paoletti P (2009) Mechanism of differential control of NMDA receptor activity by NR2 subunits. Nature 459(7247):703–707. https://doi.org/10.1038/nature07993
Gupta A, Devi LA, Gomes I (2011) Potentiation of m-opioid receptor-mediated signaling by ketamine. J Neurochem 119(2):294–302. https://doi.org/10.1111/j.1471-4159.2011.07361.x
Gupta SC, Ravikrishnan A, Liu J, Mao Z, Pavuluri R, Hillman BG, Gandhi PJ, Stairs DJ, Li M, Ugale RR, Monaghan DT, Dravid SM (2016) The NMDA receptor GluN2C subunit controls cortical excitatory-inhibitory balance, neuronal oscillations and cognitive function. Sci Rep 6:38321. https://doi.org/10.1038/srep38321
Hagino Y, Kasai S, Han W, Yamamoto H, Nabeshima T, Mishina M, Ikeda K (2010) Essential role of NMDA receptor channel e4 subunit (GluN2D) in the effects of phencyclidine, but not methamphetamine. PLoS One 5(10):e13722. https://doi.org/10.1371/journal.pone.0013722
Han LC, Yao LN, Wu SX, Yang YH, Xu LX, Chai W (2010) The effect of ketamine on N-methyl-D-aspartate receptor subunit expression in neonatal rats. Eur J Anaesthesiol 27(2):181–186. https://doi.org/10.1097/EJA.0b013e328330d453
Hardingham GE, Bading H (2010) Synaptic versus extrasynaptic NMDA receptor signalling: implications for neurodegenerative disorders. Nat Rev Neurosci 11(10):682–696. https://doi.org/10.1038/nrn2911
Higgins GA, Ballard TM, Enderlin M, Haman M, Kemp JA (2005) Evidence for improved performance in cognitive tasks following selective NR2B NMDA receptor antagonist pre-treatment in the rat. Psychopharmacology (Berl) 179(1):85–98. https://doi.org/10.1007/s00213-005-2203-9
Hikosaka O (2010) The habenula: from stress evasion to value-based decision-making. Nat Rev Neurosci 11(7):503–513. https://doi.org/10.1038/nrn2866
Hillman BG, Gupta SC, Stairs DJ, Buonanno A, Dravid SM (2011) Behavioral analysis of NR2C knockout mouse reveals deficit in acquisition of conditioned fear and working memory. Neurobiol Learn Mem 95(4):404–414. https://doi.org/10.1016/j.nlm.2011.01.008
Homayoun H, Moghaddam B (2007) NMDA receptor hypofunction produces opposite effects on prefrontal cortex interneurons and pyramidal neurons. J Neurosci 27(43):11496–11500. https://doi.org/10.1523/JNEUROSCI.2213-07.2007

Ide S, Ikeda K (2018) Mechanisms of the antidepressant effects of ketamine enantiomers and their metabolites. Biol Psychiatry 84(8):551–552. https://doi.org/10.1016/j.biopsych.2018.07.018

Ide S, Ikekubo Y, Mishina M, Hashimoto K, Ikeda K (2017) Role of NMDA receptor GluN2D subunit in the antidepressant effects of enantiomers of ketamine. J Pharmacol Sci 135(3):138–140. https://doi.org/10.1016/j.jphs.2017.11.001

Ide S, Ikekubo Y, Mishina M, Hashimoto K, Ikeda K (2019) Cognitive impairment that is induced by (*R*)-ketamine is abolished in NMDA GluN2D receptor subunit knockout mice. Int J Neuropsychopharmacol 22(7):449–452. https://doi.org/10.1093/ijnp/pyz025

Insel TR, Wang PS (2009) The STAR∗D trial: revealing the need for better treatments. Psychiatr Serv 60(11):1466–1467. https://doi.org/10.1176/appi.ps.60.11.1466

Kantrowitz JT, Javitt DC (2010) N-methyl-D-aspartate (NMDA) receptor dysfunction or dysregulation: the final common pathway on the road to schizophrenia? Brain Res Bull 83(3-4):108–121. https://doi.org/10.1016/j.brainresbull.2010.04.006

Kapur S, Seeman P (2002) NMDA receptor antagonists ketamine and PCP have direct effects on the dopamine D2 and serotonin 5-HT2 receptors: implications for models of schizophrenia. Mol Psychiatry 7(8):837–844. https://doi.org/10.1038/sj.mp.4001093

Kessler RC, Berglund P, Demler O, Jin R, Koretz D, Merikangas KR, Rush AJ, Walters EE, Wang PS (2003) The epidemiology of major depressive disorder: results from the National Comorbidity Survey Replication (NCS-R). JAMA 289(23):3095–3105. https://doi.org/10.1001/jama.289.23.3095

Khlestova E, Johnson JW, Krystal JH, Lisman J (2016) The role of GluN2C-containing NMDA receptors in ketamine's psychotogenic action and in schizophrenia models. J Neurosci 36(44):11151–11157. https://doi.org/10.1523/JNEUROSCI.1203-16.2016

Klepstad P, Maurset A, Moberg ER, Oye I (1990) Evidence of a role for NMDA receptors in pain perception. Eur J Pharmacol 187(3):513–518

Kocsis B (2012) Differential role of NR2A and NR2B subunits in N-methyl-D-aspartate receptor antagonist-induced aberrant cortical gamma oscillations. Biol Psychiatry 71(11):987–995. https://doi.org/10.1016/j.biopsych.2011.10.002

Krystal JH, Sanacora G, Duman RS (2013) Rapid-acting glutamatergic antidepressants: the path to ketamine and beyond. Biol Psychiatry 73(12):1133–1141. https://doi.org/10.1016/j.biopsych.2013.03.026

Kutsuwada T, Sakimura K, Manabe T, Takayama C, Katakura N, Kushiya E, Natsume R, Watanabe M, Inoue Y, Yagi T, Aizawa S, Arakawa M, Takahashi T, Nakamura Y, Mori H, Mishina M (1996) Impairment of suckling response, trigeminal neuronal pattern formation, and hippocampal LTD in NMDA receptor ε2 subunit mutant mice. Neuron 16(2):333–344

Li N, Lee B, Liu RJ, Banasr M, Dwyer JM, Iwata M, Li XY, Aghajanian G, Duman RS (2010) mTOR-dependent synapse formation underlies the rapid antidepressant effects of NMDA antagonists. Science 329(5994):959–964. https://doi.org/10.1126/science.1190287

Li N, Liu RJ, Dwyer JM, Banasr M, Lee B, Son H, Li XY, Aghajanian G, Duman RS (2011) Glutamate N-methyl-D-aspartate receptor antagonists rapidly reverse behavioral and synaptic deficits caused by chronic stress exposure. Biol Psychiatry 69(8):754–761. https://doi.org/10.1016/j.biopsych.2010.12.015

Lima-Ojeda JM, Vogt MA, Pfeiffer N, Dormann C, Kohr G, Sprengel R, Gass P, Inta D (2013) Pharmacological blockade of GluN2B-containing NMDA receptors induces antidepressant-like effects lacking psychotomimetic action and neurotoxicity in the perinatal and adult rodent brain. Prog Neuropsychopharmacol Biol Psychiatry 45:28–33. https://doi.org/10.1016/j.pnpbp.2013.04.017

Maeng S, Zarate CA Jr, Du J, Schloesser RJ, McCammon J, Chen G, Manji HK (2008) Cellular mechanisms underlying the antidepressant effects of ketamine: role of α-amino-3-hydroxy-5-methylisoxazole-4-propionic acid receptors. Biol Psychiatry 63(4):349–352. https://doi.org/10.1016/j.biopsych.2007.05.028

Mallick F, McCullumsmith CB (2016) Ketamine for treatment of suicidal ideation and reduction of risk for suicidal behavior. Curr Psychiatry Rep 18(6):61. https://doi.org/10.1007/s11920-016-0680-7

Martucci L, Wong AH, De Luca V, Likhodi O, Wong GW, King N, Kennedy JL (2006) N-methyl-D-aspartate receptor NR2B subunit gene GRIN2B in schizophrenia and bipolar disorder: polymorphisms and mRNA levels. Schizophr Res 84(2-3):214–221. https://doi.org/10.1016/j.schres.2006.02.001

McNally JM, McCarley RW, McKenna JT, Yanagawa Y, Brown RE (2011) Complex receptor mediation of acute ketamine application on in vitro gamma oscillations in mouse prefrontal cortex: modeling gamma band oscillation abnormalities in schizophrenia. Neuroscience 199:51–63. https://doi.org/10.1016/j.neuroscience.2011.10.015

Miller OH, Yang L, Wang CC, Hargroder EA, Zhang Y, Delpire E, Hall BJ (2014) GluN2B-containing NMDA receptors regulate depression-like behavior and are critical for the rapid antidepressant actions of ketamine. Elife 3:e03581. https://doi.org/10.7554/eLife.03581

Miyamoto Y, Yamada K, Noda Y, Mori H, Mishina M, Nabeshima T (2001) Hyperfunction of dopaminergic and serotonergic neuronal systems in mice lacking the NMDA receptor ε1 subunit. J Neurosci 21(2):750–757

Moghaddam B, Adams B, Verma A, Daly D (1997) Activation of glutamatergic neurotransmission by ketamine: a novel step in the pathway from NMDA receptor blockade to dopaminergic and cognitive disruptions associated with the prefrontal cortex. J Neurosci 17(8):2921–2927

Monyer H, Burnashev N, Laurie DJ, Sakmann B, Seeburg PH (1994) Developmental and regional expression in the rat brain and functional properties of four NMDA receptors. Neuron 12(3):529–540

Morgan CJ, Curran HV (2012) Ketamine use: a review. Addiction 107(1):27–38. https://doi.org/10.1111/j.1360-0443.2011.03576.x

Murray TF, Leid ME (1984) Interaction of dextrorotatory opioids with phencyclidine recognition sites in rat brain membranes. Life Sci 34(20):1899–1911

Neske GT, Patrick SL, Connors BW (2015) Contributions of diverse excitatory and inhibitory neurons to recurrent network activity in cerebral cortex. J Neurosci 35(3):1089–1105. https://doi.org/10.1523/JNEUROSCI.2279-14.2015

Nosyreva E, Szabla K, Autry AE, Ryazanov AG, Monteggia LM, Kavalali ET (2013) Acute suppression of spontaneous neurotransmission drives synaptic potentiation. J Neurosci 33(16):6990–7002. https://doi.org/10.1523/JNEUROSCI.4998-12.2013

Paoletti P, Bellone C, Zhou Q (2013) NMDA receptor subunit diversity: impact on receptor properties, synaptic plasticity and disease. Nat Rev Neurosci 14(6):383–400. https://doi.org/10.1038/nrn3504

Petrenko AB, Yamakura T, Fujiwara N, Askalany AR, Baba H, Sakimura K (2004) Reduced sensitivity to ketamine and pentobarbital in mice lacking the N-methyl-D-aspartate receptor GluRε1 subunit. Anesth Analg 99(4):1136–1140. https://doi.org/10.1213/01.ANE.0000131729.54986.30

Petrenko AB, Yamakura T, Sakimura K, Baba H (2014) Defining the role of NMDA receptors in anesthesia: are we there yet? Eur J Pharmacol 723:29–37. https://doi.org/10.1016/j.ejphar.2013.11.039

Picard N, Takesian AE, Fagiolini M, Hensch TK (2019) NMDA 2A receptors in parvalbumin cells mediate sex-specific rapid ketamine response on cortical activity. Mol Psychiatry 24(6):828–838. https://doi.org/10.1038/s41380-018-0341-9

Preskorn SH, Baker B, Kolluri S, Menniti FS, Krams M, Landen JW (2008) An innovative design to establish proof of concept of the antidepressant effects of the NR2B subunit selective N-methyl-D-aspartate antagonist, CP-101,606, in patients with treatment-refractory major depressive disorder. J Clin Psychopharmacol 28(6):631–637. https://doi.org/10.1097/JCP.0b013e31818a6cea

Price RB, Nock MK, Charney DS, Mathew SJ (2009) Effects of intravenous ketamine on explicit and implicit measures of suicidality in treatment-resistant depression. Biol Psychiatry 66(5):522–526. https://doi.org/10.1016/j.biopsych.2009.04.029

Rampon C, Tang YP, Goodhouse J, Shimizu E, Kyin M, Tsien JZ (2000) Enrichment induces structural changes and recovery from nonspatial memory deficits in CA1 NMDAR1-knockout mice. Nat Neurosci 3(3):238–244. https://doi.org/10.1038/72945

Robson MJ, Elliott M, Seminerio MJ, Matsumoto RR (2012) Evaluation of sigma (s) receptors in the antidepressant-like effects of ketamine in vitro and in vivo. Eur Neuropsychopharmacol 22(4):308–317. https://doi.org/10.1016/j.euroneuro.2011.08.002

Rothstein JD, Dykes-Hoberg M, Pardo CA, Bristol LA, Jin L, Kuncl RW, Kanai Y, Hediger MA, Wang Y, Schielke JP, Welty DF (1996) Knockout of glutamate transporters reveals a major role for astroglial transport in excitotoxicity and clearance of glutamate. Neuron 16(3):675–686

Sapkota K, Mao Z, Synowicki P, Lieber D, Liu M, Ikezu T, Gautam V, Monaghan DT (2016) GluN2D N-methyl-D-aspartate receptor subunit contribution to the stimulation of brain activity and gamma oscillations by ketamine: implications for schizophrenia. J Pharmacol Exp Ther 356(3):702–711. https://doi.org/10.1124/jpet.115.230391

Sato Y, Kobayashi E, Murayama T, Mishina M, Seo N (2005) Effect of N-methyl-D-aspartate receptor ε1 subunit gene disruption of the action of general anesthetic drugs in mice. Anesthesiology 102(3):557–561

Seamans J (2008) Losing inhibition with ketamine. Nat Chem Biol 4(2):91–93. https://doi.org/10.1038/nchembio0208-91

Shi Q, Guo L, Patterson TA, Dial S, Li Q, Sadovova N, Zhang X, Hanig JP, Paule MG, Slikker W Jr, Wang C (2010) Gene expression profiling in the developing rat brain exposed to ketamine. Neuroscience 166(3):852–863. https://doi.org/10.1016/j.neuroscience.2010.01.007

Singh JB, Fedgchin M, Daly E, Xi L, Melman C, De Bruecker G, Tadic A, Sienaert P, Wiegand F, Manji H, Drevets WC, Van Nueten L (2016) Intravenous esketamine in adult treatment-resistant depression: a double-blind, double-randomization, placebo-controlled study. Biol Psychiatry 80(6):424–431. https://doi.org/10.1016/j.biopsych.2015.10.018

Sutton MA, Taylor AM, Ito HT, Pham A, Schuman EM (2007) Postsynaptic decoding of neural activity: eEF2 as a biochemical sensor coupling miniature synaptic transmission to local protein synthesis. Neuron 55(4):648–661. https://doi.org/10.1016/j.neuron.2007.07.030

Swanger SA, Vance KM, Pare JF, Sotty F, Fog K, Smith Y, Traynelis SF (2015) NMDA receptors containing the GluN2D subunit control neuronal function in the subthalamic nucleus. J Neurosci 35(48):15971–15983. https://doi.org/10.1523/JNEUROSCI.1702-15.2015

Tozzi A, de Iure A, Bagetta V, Tantucci M, Durante V, Quiroga-Varela A, Costa C, Di Filippo M, Ghiglieri V, Latagliata EC, Wegrzynowicz M, Decressac M, Giampa C, Dalley JW, Xia J, Gardoni F, Mellone M, El-Agnaf OM, Ardah MT, Puglisi-Allegra S, Bjorklund A, Spillantini MG, Picconi B, Calabresi P (2016) Alpha-synuclein produces early behavioral alterations via striatal cholinergic synaptic dysfunction by interacting with GluN2D N-methyl-D-aspartate receptor subunit. Biol Psychiatry 79(5):402–414. https://doi.org/10.1016/j.biopsych.2015.08.013

Traynelis SF, Wollmuth LP, McBain CJ, Menniti FS, Vance KM, Ogden KK, Hansen KB, Yuan H, Myers SJ, Dingledine R (2010) Glutamate receptor ion channels: structure, regulation, and function. Pharmacol Rev 62(3):405–496. https://doi.org/10.1124/pr.109.002451

Watanabe M, Inoue Y, Sakimura K, Mishina M (1992) Developmental changes in distribution of NMDA receptor channel subunit mRNAs. Neuroreport 3(12):1138–1140

Williams NR, Heifets BD, Blasey C, Sudheimer K, Pannu J, Pankow H, Hawkins J, Birnbaum J, Lyons DM, Rodriguez CI, Schatzberg AF (2018) Attenuation of antidepressant effects of ketamine by opioid receptor antagonism. Am J Psychiatry 175(12):1205–1215. https://doi.org/10.1176/appi.ajp.2018.18020138

Winters BD, Bussey TJ (2005) Glutamate receptors in perirhinal cortex mediate encoding, retrieval, and consolidation of object recognition memory. J Neurosci 25(17):4243–4251. https://doi.org/10.1523/JNEUROSCI.0480-05.2005

World Health Organization (2017) The selection and use of essential medicines, 21st edn. World Health Organization, Geneva

Wyllie DJ, Livesey MR, Hardingham GE (2013) Influence of GluN2 subunit identity on NMDA receptor function. Neuropharmacology 74:4–17. https://doi.org/10.1016/j.neuropharm.2013.01.016

Yamamoto H, Kamegaya E, Sawada W, Hasegawa R, Yamamoto T, Hagino Y, Takamatsu Y, Imai K, Koga H, Mishina M, Ikeda K (2013) Involvement of the N-methyl-D-aspartate receptor GluN2D subunit in phencyclidine-induced motor impairment, gene expression, and increased Fos immunoreactivity. Mol Brain 6:56. https://doi.org/10.1186/1756-6606-6-56

Yamamoto T, Nakayama T, Yamaguchi J, Matsuzawa M, Mishina M, Ikeda K, Yamamoto H (2016) Role of the NMDA receptor GluN2D subunit in the expression of ketamine-induced behavioral sensitization and region-specific activation of neuronal nitric oxide synthase. Neurosci Lett 610:48–53. https://doi.org/10.1016/j.neulet.2015.10.049

Yamamoto H, Kamegaya E, Hagino Y, Takamatsu Y, Sawada W, Matsuzawa M, Ide S, Yamamoto T, Mishina M, Ikeda K (2017) Loss of GluN2D subunit results in social recognition deficit, social stress, 5-HT2C receptor dysfunction, and anhedonia in mice. Neuropharmacology 112(Pt A):188–197. https://doi.org/10.1016/j.neuropharm.2016.07.036

Yang C, Shirayama Y, Zhang JC, Ren Q, Yao W, Ma M, Dong C, Hashimoto K (2015) R-ketamine: a rapid-onset and sustained antidepressant without psychotomimetic side effects. Transl Psychiatry 5:e632. https://doi.org/10.1038/tp.2015.136

Yang C, Han M, Zhang JC, Ren Q, Hashimoto K (2016) Loss of parvalbumin-immunoreactivity in mouse brain regions after repeated intermittent administration of esketamine, but not R-ketamine. Psychiatry Res 239:281–283. https://doi.org/10.1016/j.psychres.2016.03.034

Yang Y, Cui Y, Sang K, Dong Y, Ni Z, Ma S, Hu H (2018) Ketamine blocks bursting in the lateral habenula to rapidly relieve depression. Nature 554(7692):317–322. https://doi.org/10.1038/nature25509

Yuan H, Hansen KB, Vance KM, Ogden KK, Traynelis SF (2009) Control of NMDA receptor function by the NR2 subunit amino-terminal domain. J Neurosci 29(39):12045–12058. https://doi.org/10.1523/JNEUROSCI.1365-09.2009

Zanos P, Gould TD (2018) Mechanisms of ketamine action as an antidepressant. Mol Psychiatry 23(4):801–811. https://doi.org/10.1038/mp.2017.255

Zanos P, Moaddel R, Morris PJ, Georgiou P, Fischell J, Elmer GI, Alkondon M, Yuan P, Pribut HJ, Singh NS, Dossou KS, Fang Y, Huang XP, Mayo CL, Wainer IW, Albuquerque EX, Thompson SM, Thomas CJ, Zarate CA Jr, Gould TD (2016) NMDAR inhibition-independent antidepressant actions of ketamine metabolites. Nature 533(7604):481–486. https://doi.org/10.1038/nature17998

Zhang X, Chergui K (2015) Dopamine depletion of the striatum causes a cell-type specific reorganization of GluN2B- and GluN2D-containing NMDA receptors. Neuropharmacology 92:108–115. https://doi.org/10.1016/j.neuropharm.2015.01.007

Zhang C, Li Z, Wu Z, Chen J, Wang Z, Peng D, Hong W, Yuan C, Wang Z, Yu S, Xu Y, Xu L, Xiao Z, Fang Y (2014a) A study of N-methyl-D-aspartate receptor gene (GRIN2B) variants as predictors of treatment-resistant major depression. Psychopharmacology (Berl) 231(4):685–693. https://doi.org/10.1007/s00213-013-3297-0

Zhang JC, Li SX, Hashimoto K (2014b) R(-)-ketamine shows greater potency and longer lasting antidepressant effects than S(+)-ketamine. Pharmacol Biochem Behav 116:137–141. https://doi.org/10.1016/j.pbb.2013.11.033

Zhang X, Feng ZJ, Chergui K (2014c) Allosteric modulation of GluN2C/GluN2D-containing NMDA receptors bidirectionally modulates dopamine release: implication for Parkinson's disease. Br J Pharmacol 171(16):3938–3945. https://doi.org/10.1111/bph.12758

Effects of Adjunctive Ketamine Intravenous Infusion in Taiwanese Patients with Treatment-Resistant Depression: Antidepression, Antisuicidality, BDNF Val66Met, and Brain Imaging

Mu-Hong Chen and Tung-Ping Su

Abstract About 21% of Taiwanese patients with major depression developed treatment-resistant depression (TRD) during 1-year follow-up in Taiwan. TRD was commonly coupled with functional impairment, poor quality of life, suicide ideation and attempts, self-injurious behaviors, and a high relapse rate. From 2012 and 2015, we had conducted the first randomized, double-blind, placebo control trial in Asian countries to examine the therapeutic efficacy of a single low-dose ketamine infusion in Taiwanese patients with TRD. Seventy one subjects were evenly distributed in three dose groups with 0.5 mg/kg, 0.2 mg/kg, and placebo, respectively. Responder was identified by response (≥50% reduction of mood ratings) at any two daily HAMD measures during the period of 24–96 h (day 2–5) post-ketamine infusion. There was a significant difference in the response rate across the three groups (0.5 mg/kg: 45.8%; 0.2 mg/kg: 39.1%; placebo: 12.5%; $p = 0.03$), which is much lower than that in the Caucasians (70%). Two factors might be related: lower serum ketamine levels and lower Val/Val allele percentage in BDNF Val66Met genotyping found in the Taiwanese patients. In addition to the rapid antidepressant effect of ketamine, a greater antisuicidal effect (59%) was also identified. The former may only account for 52.7% of the latter, indicating that the antisuicidality may be independent from antidepressiveness. Single dose of ketamine only resulted in short-lived

M.-H. Chen
Division of Psychiatry, Faculty of Medicine, National Yang-Ming University, Taipei, Taiwan

Department of Psychiatry, Taipei Veterans General Hospital, Taipei, Taiwan

Institute of Brain Science, National Yang-Ming University, Taipei, Taiwan

T.-P. Su (✉)
Division of Psychiatry, Faculty of Medicine, National Yang-Ming University, Taipei, Taiwan

Department of Psychiatry, Taipei Veterans General Hospital, Taipei, Taiwan

Institute of Brain Science, National Yang-Ming University, Taipei, Taiwan

Department of Medical Research, Taipei Veterans General Hospital, Taipei, Taiwan

Department of Psychiatry, Cheng Hsin General Hospital, Taipei, Taiwan

K. Hashimoto et al. (eds.), *Ketamine*,
https://doi.org/10.1007/978-981-15-2902-3_11

psychedelic/dissociation adverse effect, which was resolved within 2 h post-infusion. Up to present, no long-term side effects were identified, given a single-dose administration. Using ^{18}F-FDG-PET scanning and 24 h post-ketamine infusion, glucose metabolism of the prefrontal cortex (PFC), supplementary motor area (SMA), and dorsal anterior cingulate cortex (dACC) were activated immediately (40 min post-infusion), and activation of SMA and dACC could sustain for 1 day, which may contribute to the persistent antidepressant effect of ketamine beyond its half-life, suggesting that a short activation in the PFC engendered by ketamine infusion may be a kindler, facilitating the persistent increase in glucose metabolism in the SMA and dACC; therefore, the PFC may be still considered to play a key role in improving TRD. These findings were also supported by a simple wearable forehead EEG monitoring from baseline to 40 min post-infusion, revealing that ketamine may increase the theta and low alpha power and decrease asymmetry in the PFC. Finally, a maintenance trial in a double-blind, randomized fashion using a partial NMDA agonist, D-cycloserine (DCS) vs. placebo, was conducted in the ketamine responders. This DCS augmentation treatment was not superior to placebo in maintaining the initial antidepressant response to ketamine infusion, but DCS did appear to maintain the antisuicidal effect during the 6-week follow-up study.

Keywords Ketamine · Depression · Suicide · BDNF · Brain imaging

1 Introduction

Major depressive disorder (MDD) is a severe chronic mental disorder that was predicted to be the leading cause of the disease burden by 2015 (WHO 2008). The Sequenced Treatment Alternatives to Relieve Depression (STAR∗D) study determined that up to 40% of MDD patients did not reach symptomatic remission after at least two trials of antidepressants (Howland 2008). Such patients were then defined as having treatment-resistant depression (TRD), and approximately 30% of MDD patients were found to continue to suffer from depression and related symptoms, such as insomnia, psychomotor retardation, and suicidal ideation, after four trials of different antidepressant treatments, including combination therapy and augmentation therapy (Howland 2008).

In Taiwan, the prevalence of common mental disorders doubled from 11.5% in 1990 to 23.8% in 2010 in parallel with the increase in national rates of unemployment, divorce, and suicide (Fu et al. 2013). Based on a cohort of 704,265 adults randomly sampled from Taiwan National Health Insurance Research database, among 2751 patients with MDD who were treated with antidepressants, approximately 21% ($n = 576$) developed TRD, which was defined as the failure to respond to more than two adequate antidepressant trials, during the 1-year follow-up (Fife et al. 2017). TRD was commonly coupled with functional impairment, poor quality of life, suicide ideation and attempts, self-injurious behaviors, and a high relapse rate (Al-Harbi 2012).

Between 2012 and 2015, we had conducted the first randomized, double-blind, placebo control trial, approved by the Institutional Review Board of Taipei Veterans General Hospital (VGHTPE) and the Department of Health of Taiwan in Asian countries and in Taiwan to examine the therapeutic efficacy of a single low-dose ketamine infusion in Taiwanese patients with TRD. Here, we summarized our study findings and proposed our personal clinical experiences for the application of ketamine infusion in TRD treatment in Asian and Taiwanese population.

2 Clinical Response of a Single Low-Dose Ketamine Infusion in Taiwanese Patients with TRD

Based on the hypothesis that Asian patients with MDD may respond fairly to the lower dose of antidepressants compared with Caucasian patients (Hong Ng et al. 2006; Lin and Shen 1991), Taiwanese patients with TRD were randomized to receive an infusion of saline (placebo group) or R/S-ketamine hydrochloride (Ketalar, Pfizer Pharmacueticals) at doses of 0.2 mg/kg (lower dose group) or 0.5 mg/kg (standard dose group) in our study (UMIN Clinical Trials Registry: UMIN000016985) (Su et al. 2017a). Hamilton Depression Rating Scale (HAMD) and Montgomery-Asberg Depression Rating Scale (MADRS) were administered in person prior to the initiation of test infusions and 40, 80, 120, and 240 min later, and telephone ratings were conducted 24, 48, 72, 96, 120, 144, and 288 h post-infusion.

2.1 *Treatment Response and Ketamine Infusion in Taiwanese Patients with TRD*

TRD patients treated with 0.5 mg/kg ketamine infusion exhibited more HAMD score reduction than those with 0.2 mg/kg ketamine infusion and normal saline (Fig. 1). Responder status was identified by response (≥50% reduction of mood ratings) at any two daily HAMD measures during the period of 24–96 h (day 2 to 5) post-infusion. There was a significant difference in the response rate across the three groups (0.5 mg/kg: 45.8%; 0.2 mg/kg: 39.1%; placebo: 12.5%; $p = 0.03$) with a significant linear trend test for the dose effect ($p = 0.01$). Post hoc test analyses indicated that responder rate was greater in the 0.5 mg/kg than placebo ($p = 0.01$) and in the 0.2 mg/kg than placebo ($p = 0.05$) but not between the two ketamine groups ($p = 0.77$). However, several studies with a similar study design investigating the therapeutic efficacy of 0.5 mg/kg ketamine in Caucasian patients with TRD have demonstrated that the response rate in patients is as high as 70% (Murrough et al.

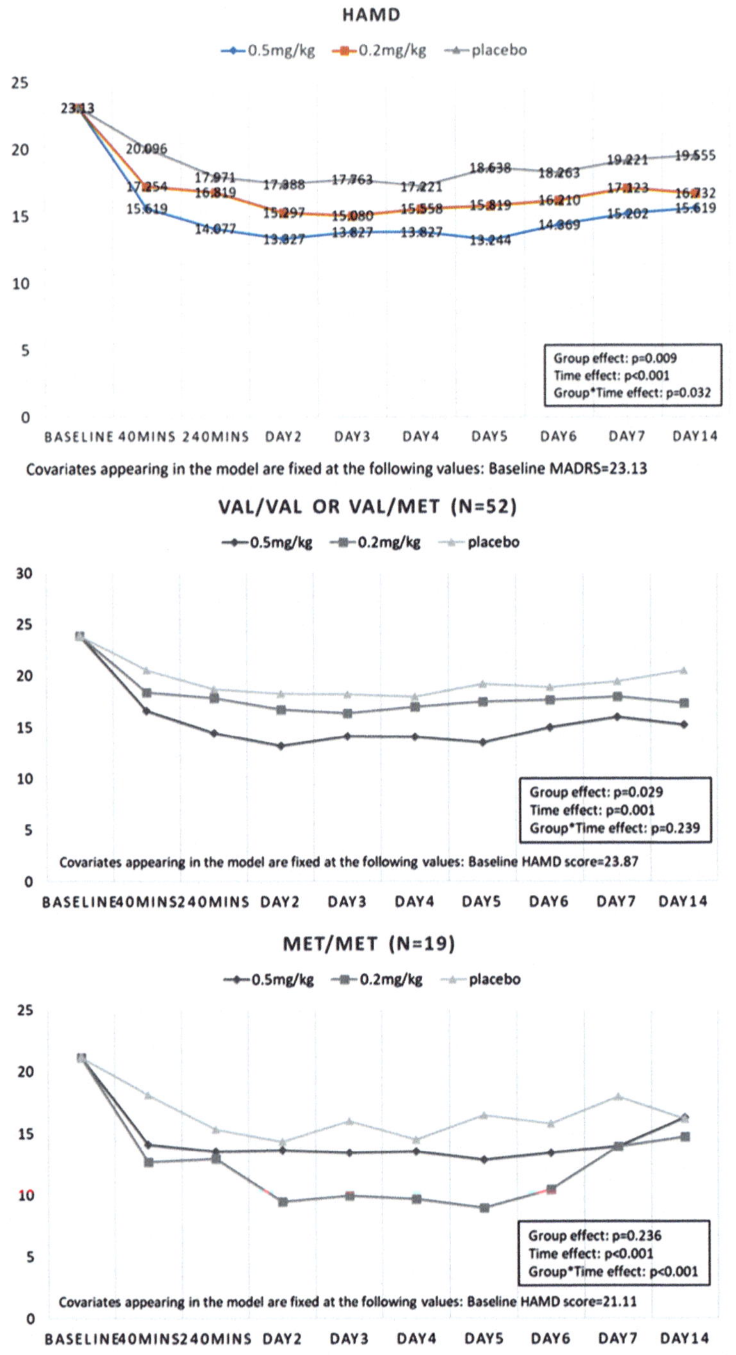

Fig. 1 Trajectory of HAMD after ketamine infusion. *HAMD* Hamilton depression rating scale

2013; Wan et al. 2015; Zarate et al. 2012). However, we only found approximately 50% response rate in Taiwanese patients receiving 0.5 mg/kg ketamine infusion, which was much lower compared with response rate in Caucasian patients (Su et al. 2017a). The lower responsiveness was attributed to the following factors: lower serum ketamine/norketamine levels and different distribution of genotypes of the BDNF Val66Met polymorphism observed in the Taiwanese subjects.

2.2 *Serum Levels of Ketamine and Norketamine After Ketamine Infusion in Taiwanese Patients with TRD*

Ketamine and norketamine levels were dose related (0.5 mg/kg group >0.2 mg/kg group) in our study. They varied by time point, but there were no significant differences by responder status. However, we found that both ketamine (115.86 ng/mL vs. 204.13 ng/mg) and norketamine (33.39 ng/mL vs. 55.52 ng/mL) levels at 40 min after ketamine infusion were much lower in Taiwanese patients than in Caucasian patients, which may confound the treatment efficacy of ketamine in Taiwanese patients (Fig. 2) (Su et al. 2017a; Zarate et al. 2012). However, the mechanisms of the lower ketamine and norketamine levels at post-infusion remained unknown in Taiwanese patients with TRD. The pharmacokinetics and pharmacodynamics of ketamine infusion need further investigation between Asian and Caucasian population.

2.3 *BDNF and Treatment Response of Ketamine Infusion in Taiwanese Patients with TRD*

The allele distribution of Val66Met BDNF polymorphism significantly varies across ethnicities, with the derived Met allele of Val66Met ranging in frequency from 0 to 72% across populations (Petryshen et al. 2010). Estimation of the Pooled Prevalence

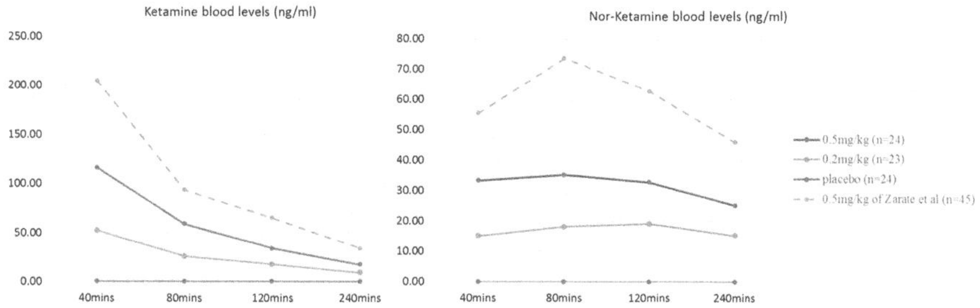

Fig. 2 Levels of ketamine and norketamine across time points

of the BDNF Val66Met Allele found that the frequency of Val allele was approximately 80% in Caucasians, but was much lower (<50%) in Asians, which may partially explain the difference of prevalence and susceptibility of mental disorders between Caucasians and Asians (Gratacos et al. 2007).

In our study, the distribution of genotypes of the Val66Met BDNF polymorphism was Val/Val (17%), Val/Met (56.3%), and Met/Met (26.8%) (Su et al. 2017a). There were no significant differences in ketamine treatment response between carriers of the Met allele and Val/Val patients (responder rate: 33.9% vs. 25.0%, $p = 0.55$). Despite Val66Met BDNF is not a biomarker of treatment response to ketamine infusion, we found that low-dose ketamine infusion was only effective for TRD patients with Val/Val or Val/Met in our study. As for those with Met/Met, the treatment efficacy of low-dose ketamine infusion did not differ between groups. However, the non-effectiveness of ketamine infusion may be due to limited cases with Met/Met (Fig. 1). A higher dose (i.e., 0.8–1 mg/kg) of ketamine infusion may be necessary for patients with TRD carrying Met/Met.

2.4 BDNF and Antisuicidal Effect of Ketamine Infusion in Taiwanese Patients with TRD

Traditional antidepressants, such as selective serotonin reuptake inhibitors and serotonin norepinephrine reuptake inhibitors, require 2–3 weeks to achieve antidepressant efficacy. The period between the initiation of antidepressant use and their optimal efficacy is a crucial period for treatment as well as depression-related consequences, such as suicide. During this period, the risk of suicide remains high; therefore, choosing antidepressants with rapid efficacy that reduce suicide risk is crucial in clinical practice and the mental health domain.

A meta-analysis of 10 randomized placebo control trials involving 167 patients with MDD, bipolar depression, or posttraumatic stress disorder reported that ketamine rapidly (within 1 day) and significantly reduced suicidal ideation in terms of both clinician-administered and self-report outcome measures (Wilkinson et al. 2018). However, the duration of the antisuicidal effects of a single low-dose ketamine infusion may not be longer than a week based on previous western studies (Murrough et al. 2015).

In our study, patients with TRD that received 0.5 mg/kg ketamine infusion exhibited a significantly lower score in item 3 of the HAMD and in item 10 of MADRS than the groups that received 0.2 mg/kg ketamine or placebo infusion (Chen et al. 2019). Based on item 10 score of MADRS, among those carrying any Val allele of BDNF, both 0.5 and 0.2 mg/kg ketamine infusions were effective in the reduction in suicidal thoughts, but among those with Met/Met of BDNF, only 0.5 mg/kg ketamine infusion was effective in the reduction in suicidal thoughts (Chen et al. 2019). Furthermore, for the higher severity of suicidal thought (item 3 of the HAMD ≥2 [wishes he or she were dead or any thoughts of possible death to self] or item 10 of the MADRS ≥4 [suicidal thoughts are common, and suicide is considered as a possible solution, but without specific plans or intention]), we found that 0.5 mg/kg ketamine infusion was more effective in reducing suicidal symptoms than 0.2 mg/kg and placebo infusion (Chen et al. 2019). Our findings may suggest a higher dose of ketamine infusion for Taiwanese patients with TRD carrying Met/Met BDNF polymorphism and those having the higher severity of suicidal thought.

2.5 Greater Antisuicidal Than Antidepressant Effect of Ketamine Infusion in Taiwanese Patients with TRD

Reanalysis of our study data (Su et al. 2017a) showed a robust and fast antidepressant and antisuicidal effect with significant reduction of ratings by high dose (0.5 mg/kg) vs. placebo from 40-min to day 7 after ketamine infusion. Symptom attenuation in HAMD_17 rating within 1 week post-infusion was not only seen in depression (43%) but even greater in suicidal ideation (59%) (Su et al. 2017b). Although there were correlations for these two symptom changes ($r = 0.76$, $p < 0.05$), regression analysis elicited decrease percentage at 240 min accounted only for 52.7% of reduction in suicidal ideation, suggesting that improvements of suicidal ideation after ketamine infusion are related to, but not completely driven by, improvement in depression. There might have other neural substrate influenced by ketamine to account for decreasing suicidal ideation.

2.6 Adverse Effects of Low-Dose Ketamine Infusion in Taiwanese Patients with TRD

Psychotomimetic and dissociative symptoms and hypertension were the common, but temporary, adverse effects during ketamine infusion in Taiwanese patients with TRD (Su et al. 2017a; Zarate et al. 2012). They always resolved completely within 80–120 min post-infusion. Nausea was another adverse effect (approximately 10%) that clinicians should pay attention.

3 Cognitive Effect of a Single Low-Dose Ketamine Infusion in Taiwanese Patients with TRD

Cognitive impairment may be an important concern in the clinical practice when TRD is treated with low-dose ketamine infusion because animal studies have demonstrated that a single anesthetic dose and repeated subanesthetic doses of ketamine infusion can impair cognitive function and increase impulsive behaviors (Ding et al.

2016; Melo et al. 2016; Wang et al. 2014). A Taiwanese study demonstrated that the ketamine-dependent patients had significantly poorer performance than did the controls in many cognitive tasks, including verbal memory, motor speed, verbal fluency, and attention and processing speed (Wang et al. 2018).

In our clinical trial, working memory task and a go/no-go task were performed at baseline and at Day 3 and Day 14 post-ketamine infusion (Chen et al. 2018a). We found no group effect, no time effect, and no group × time interaction effect for cognitive function among baseline, Day 3, and Day 14 among the three groups, which may indicate that a low dose of ketamine infusion did not impair cognitive function in patients with TRD (Chen et al. 2018a). Subanalyses additionally demonstrated that specific cognitive improvement measured using the go/no-go task was observed only among the responders in the 0.5-mg/kg ketamine infusion group (Chen et al. 2018a). Furthermore, increasing evidence suggested that the repeated low-dose ketamine infusion may be more beneficial for the improved treatment response and the longer antidepressant effect (Singh et al. 2016). In the future, cognitive function changes after repeated ketamine infusion should be investigated for patients with TRD.

4 Central Mechanisms of the Low-Dose Ketamine Infusion in Taiwanese Patients with TRD

At low doses, ketamine is believed to preferentially bind to and inhibit NMDARs on GABA-ergic interneurons (Ide and Ikeda 2018; Zanos and Gould 2018). The process, initiated by ketamine, in GABA interneurons leading to the subsequent increase in synaptogenesis and BDNF levels has been considered a major biological mechanism of the rapid antidepressant effect of ketamine (Ide and Ikeda 2018; Zanos and Gould 2018).

Hypofrontality and hyperactivity of limbic system structures, such as the amygdala, measured according to prefrontal cortex (PFC) glucose metabolism or functional MRI have been reported to be significantly correlated with depression severity in patients with MDD and were the biosignature of TRD (Drevets et al. 2008; Li et al. 2015; Phillips et al. 2015). The anterior cingulate cortex (ACC) dysfunction is another notable biosignature of MDD and TRD. The dorsal ACC (dACC) is a critical hub that integrates the emotional and cognitive domains of emotional and behavioral regulation, thereby connecting the PFC, supplementary motor area (SMA), and limbic system (Bush et al. 2000). The dACC synergically works with the dorsolateral PFC (DLPFC) in attentional and cognitive control and in emotional regulation through the inhibition of the hyperactive limbic system and subgenual ACC (De Raedt et al. 2015; Ressler and Mayberg 2007). Furthermore, SMA plays a critical role in human volition, executive function, and integration of affective, behavioral, and cognitive functions, which were severely impaired in patients with TRD (Haggard 2008; Leisman et al. 2016; Nachev et al. 2008). Impairment in attention regulation and cognitive control function was correlated with hypoactivation in the SMA and ACC (Halari et al. 2009).

In our randomized placebo control clinical trial, all patients with TRD were assessed with ^{18}F-FDG-PET scan at baseline and 48 with PET scan immediately at 40 min post-infusion for the evaluation of the rapid onset antidepressant effect of ketamine and the other 24 with PET at 24-h post-infusion for the persistent antidepressant effect of ketamine because the half-life values of ketamine and its active metabolites, norketamine and dehydronorketamine, are approximately 3, 5, and 7 h, respectively (Chen et al. 2018b; Li et al. 2016). We found that the glucose metabolism of the PFC, SMA, and dACC in patients with TRD provided with the low-dose ketamine infusion were higher than those in the control group at 40 min post-infusion (Li et al. 2016). Interestingly, two other PET studies have not observed increased glucose metabolism in the PFC 2 h after ketamine infusion, but they also reported that the rapid antidepressant effect of ketamine infusion was mediated by activation in the dACC (Carlson et al. 2013; Lally et al. 2015). Lastly, we found that the activation in the SMA and dACC could persist 1 day after a single-dose ketamine infusion and may contribute to the persistent antidepressant effect of ketamine considerably beyond its half-life (Chen et al. 2018b). Taking the preceding pieces of evidence together, we hypothesize that the effect of ketamine infusion on PFC activation rapidly occurred within 1 h and then rapidly disappeared approximately 2 h later. A short activation in the PFC engendered by ketamine infusion may be a kindler, facilitating the persistent increase in glucose metabolism in the SMA and dACC; therefore, the PFC may still be considered to play a key role in improving TRD (Fig. 3) (Chen et al. 2018b; Li et al. 2016).

5 Identifying Ketamine Responses in Taiwanese Patients with TRD Using a Wearable Forehead EEG

Electroencephalography (EEG) has been widely used to study antidepressant treatment responses due to its broad availability and cost-effectiveness. Prefrontal EEG power at baseline was reported to predict the SSRI antidepressant response with 63% (Iosifescu et al. 2009) and 88% accuracy (Khodayari-Rostamabad et al. 2013), suggesting that the use of forehead EEG patterns might be a potential for building a baseline predictor for responses to ketamine treatment. With the development of sensor technology, an alternative to conventional EEG devices with wet electrodes and cables has emerged. Wearable wireless EEG device with dry sensors have led to a reduction in the amount of preparatory work required for a long-term monitoring and daily use (Lin et al. 2008). Since ketamine has a rapid and robust antidepressant effect in treatment-resistant depression (TRD), which is paralleling with increased glucose metabolism in the frontal area by PET-FDG, our study aim is to investigate the role of frontal EEG as predictor for clinical response to ketamine in TRD via a wireless EEG device. In our double-blind, randomized trial and placebo control

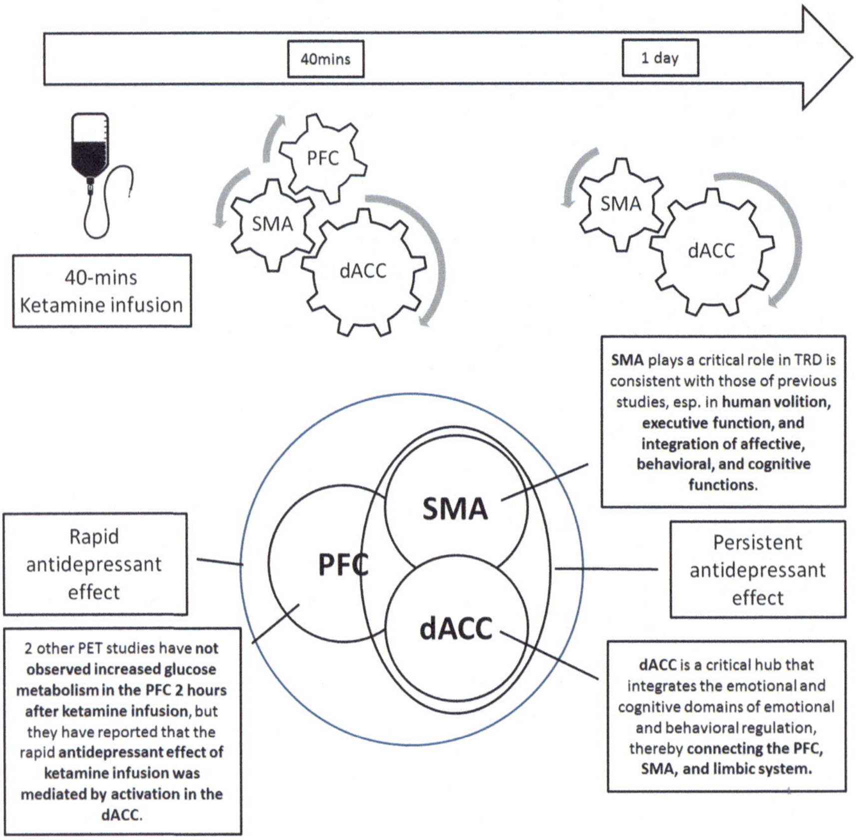

Fig. 3 Hypothesis of central mechanisms of antidepressant effect of low-dose ketamine infusion. *PFC* prefrontal cortex, *SMA* supplementary motor area, *dACC* dorsal anterior cingulate cortex

study (Cao et al. 2018), 36 patients on active ketamine dose (0.2–0.5 mg/kg) vs. 18 patients on placebo received 5-min electroencephalography (EEG) recording via a wireless EEG device with four prefrontal dry-contact sensors at baseline (0 min) and posttreatment (240 min), respectively. The ketamine responses were measured by EEG signals and Hamilton depression rating scale (HDRS) scores. Responder was identified (≥50% reduction of baseline depression symptoms at 240 min post-infusion). EEG power and hemispheric asymmetry were calculated in the delta, theta, low alpha, and high alpha bands. Our results showed that, at baseline, responders had a significantly weaker EEG theta and lower level of low alpha power than did nonresponders ($p < 0.05$). Further, in the responders, we found that ketamine increased the low relative alpha power ($p < 0.01$) and decreased its hemispheric (Fp2-Fp1) asymmetry over bilateral frontals ($p < 0.05$), which was observed in neither nonresponders nor placebo controls. The result is consistent with the previous studies that decreased frontal EEG asymmetry may be positively correlated with response to behavioral therapy in depression (Gollan et al. 2014; Jesulola et al. 2015) and is also required for the ketamine-specific fast antidepressant effect. Moreover, our baseline EEG predictor classified responders and nonresponders with 81.3 ± 9.5% accuracy, 82.1 ± 8.6% sensitivity, and 91.9 ± 7.4% specificity. In conclusion, this study provided the evidence of immediate changes of frontal activity, which may account for rapid clinical response of ketamine, and the pretreatment frontal brain activity measured by simple EEG device might be a biosignature for better outcome prediction.

6 The Maintenance of Antidepressant and Antisuicidal Effect of Low-Dose Ketamine Infusion with D-Cycloserine in Taiwanese Patients with TRD

Maintaining the initial response of ketamine infusion is the next step in therapeutic challenge in the psychiatric clinical practice (Iosifescu 2015). The safety and tolerability of repeated or long-term ketamine infusion have not been established until now; particularly, long-term ketamine misuse has been reported to cause various psychiatric and physical complications and adversities, including cognitive impairment, psychosis, and interstitial cystitis (Wang et al. 2018; Cheng et al. 2018; Chu et al. 2008).

D-Cycloserine (DCS), a partial NMDA agonist, has been reported to be potentially effective for depression augmentation treatment (Heresco-Levy et al. 2013; Henter et al. 2018). In our recent clinical trial (UMIN Clinical Trials Registry: UMIN000023581), patients with TRD (N = 32) who responded to the add-on 40-min intravenous ketamine (0.5 mg/kg) infusions at Day 1 and Day 4 in the phase 1 study were enrolled in the phase 2 double-blind randomized DCS–placebo control study (Chen, et al, 2019, accepted in Neuropsychopharmacology). In the phase 2 study, patients with TRD were randomly divided into 6-week DCS (Seromycin, Eli Lilly) treatment (250 mg/day for 2 days, 500 mg/day for 2 days, and 750 mg/day for 3 days to 1000 mg/day for 5 weeks) and placebo groups.

Unfortunately, DCS augmentation treatment was not superior to placebo in maintaining the initial antidepressant response to ketamine infusion in Taiwanese patients with TRD. But, interestingly, DCS did appear to maintain the antisuicidal effect, measured by item 3 of HAMD, of ketamine infusion during the 6-week follow-up period compared with the placebo (Fig. 4). This finding may imply that DCS augmentation was helpful for patients with TRD who responded to ketamine infusion but still had residual suicidal thought. However, the mechanisms underlying the potentially beneficial effect of DCS in depression and suicidality are not well understood and need further investigation.

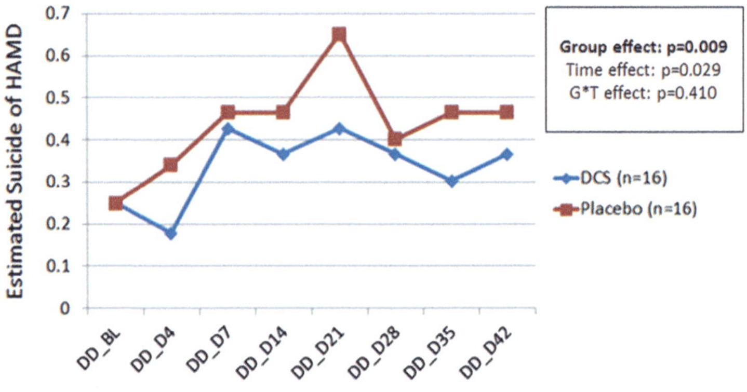

Fig. 4 Trajectory of suicide (item 3 of HAMD) in phase 2 double-blind randomized placebo control study. *HAMD* Hamilton depression rating scale, *DCS* D-cycloserine

7 Conclusion

Our clinical trials and clinical experience supported that the low-dose ketamine infusion (0.5 and 0.2 mg/kg) was a safe, tolerable, and effective treatment for Taiwanese patients with TRD although the Met allele of BDNF polymorphism is predominant in Taiwanese and Asian population. However, the higher dose (i.e., 0.8–1.0 mg/kg) of ketamine infusion may be necessary for those who carry Met/Met BDNF polymorphism or have higher severity of suicidality. We also found the potential maintenance effect for suicidality of DCS after ketamine infusions. Furthermore, the modulation of PFC-related circuits and increased activation of PFC, SMA, and dACC may play a crucial role in the antidepressant and antisuicidal effects of ketamine infusion. Further clinical trial will be required to elucidate the optimal dose (i.e., 0.5–1.0 mg/kg) and infusion frequency (once or twice per week for 2–4 weeks) of ketamine infusion for the maximized treatment outcome among Taiwanese patients with TRD. The therapeutic strategy that may maintain the initial response of ketamine infusion also needs further investigation.

Acknowledgements We express our gratitude to all patients who kindly participated in this study. We thank all research assistants, physicians, pharmacists, and nursing staffs at D020 Unit of Taipei Veterans General Hospital for their assistance during the study process, without whom this work could not have been possible. We thank Mr. I-Fan Hu for his support and friendship. The participation of Dr. Krystal was supported by the US Department of Veterans Affairs via its support for the VA National Center for PTSD, the US National Institute on Alcohol Abuse and Alcoholism (P50AA12870, M01RR00125), and the US National Center for Advancing Translational Science (UL1 RR024139). None of the aforementioned funding organizations had any role in the study design, data collection, analysis, interpretation of result, writing of the report, and the ultimate decision to submit the paper for publication.

Funding and disclosures: The study was sponsored by grants from Ministry of Science and Technology, Taiwan (101-2314-B-010-060, 102-2314-B-010-060, 107-2314-B-075-063-MY3 and 108-2314-B-075-037), Taipei Veterans General Hospital (V106B-020, V107B-010, and V107C-181), and the Kun-Po Soo Medical Foundation in Taiwan.

References

Al-Harbi KS (2012) Treatment-resistant depression: therapeutic trends, challenges, and future directions. Patient Prefer Adherence 6:369–388

Bush G, Luu P, Posner MI (2000) Cognitive and emotional influences in anterior cingulate cortex. Trends Cogn Sci 4:215–222

Cao Z, Lin CT, Ding W, Chen MH, Li CT, Su TP (2018) Identifying ketamine responses in treatment-resistant depression using a wearable forehead EEG. IEEE Trans Biomed Eng. https://doi.org/10.1109/TBME.2018.2877651

Carlson PJ, Diazgranados N, Nugent AC, Ibrahim L, Luckenbaugh DA, Brutsche N, Herscovitch P, Manji HK, Zarate CA Jr, Drevets WC (2013) Neural correlates of rapid antidepressant response to ketamine in treatment-resistant unipolar depression: a preliminary positron emission tomography study. Biol Psychiatry 73:1213–1221

Chen MH, Li CT, Lin WC, Hong CJ, Tu PC, Bai YM, Cheng CM, Su TP (2018a) Cognitive function of patients with treatment-resistant depression after a single low dose of ketamine infusion. J Affect Disord 241:1–7

Chen MH, Li CT, Lin WC, Hong CJ, Tu PC, Bai YM, Cheng CM, Su TP (2018b) Persistent antidepressant effect of low-dose ketamine and activation in the supplementary motor area and anterior cingulate cortex in treatment-resistant depression: a randomized control study. J Affect Disord 225:709–714

Chen MH, Lin WC, Wu HJ, Cheng CM, Li CT, Hong CJ, Tu PC, Bai YM, Tsai SJ, Su TP (2019) Antisuicidal effect, BDNF Val66Met polymorphism, and low-dose ketamine infusion: reanalysis of adjunctive ketamine study of Taiwanese patients with treatment-resistant depression (AKSTP-TRD). J Affect Disord 251:162–169

Cheng WJ, Chen CH, Chen CK, Huang MC, Pietrzak RH, Krystal JH, Xu K (2018) Similar psychotic and cognitive profile between ketamine dependence with persistent psychosis and schizophrenia. Schizophr Res 199:313–318

Chu PS, Ma WK, Wong SC, Chu RW, Cheng CH, Wong S, Tse JM, Lau FL, Yiu MK, Man CW (2008) The destruction of the lower urinary tract by ketamine abuse: a new syndrome? BJU Int 102:1616–1622

De Raedt R, Vanderhasselt MA, Baeken C (2015) Neurostimulation as an intervention for treatment resistant depression: from research on mechanisms towards targeted neurocognitive strategies. Clin Psychol Rev 41:61–69

Ding R, Li Y, Du A, Yu H, He B, Shen R, Zhou J, Li L, Cui W, Zhang G, Lu Y, Wu X (2016) Changes in hippocampal AMPA receptors and cognitive impairments in chronic ketamine addiction models: another understanding of ketamine CNS toxicity. Sci Rep 6:38771

Drevets WC, Price JL, Furey ML (2008) Brain structural and functional abnormalities in mood disorders: implications for neurocircuitry models of depression. Brain Struct Funct 213:93–118

Fife D, Feng Y, Wang MY, Chang CJ, Liu CY, Juang HT, Furnback W, Singh J, Wang B (2017) Epidemiology of pharmaceutically treated depression and treatment resistant depression in Taiwan. Psychiatry Res 252:277–283

Fu TS, Lee CS, Gunnell D, Lee WC, Cheng AT (2013) Changing trends in the prevalence of common mental disorders in Taiwan: a 20-year repeated cross-sectional survey. Lancet 381:235–241

Gollan JK, Hoxha D, Chihade D, Pflieger ME, Rosebrock L, Cacioppo J (2014) Frontal alpha EEG asymmetry before and after behavioral activation treat.ment for depression. Biol Psychol 99:198–208

Gratacos M, Gonzalez JR, Mercader JM, de Cid R, Urretavizcaya M, Estivill X (2007) Brain-derived neurotrophic factor Val66Met and psychiatric disorders: meta-analysis of case-control studies confirm association to substance-related disorders, eating disorders, and schizophrenia. Biol Psychiatry 61:911–922

Haggard P (2008) Human volition: towards a neuroscience of will. Nat Rev Neurosci 9:934–946

Halari R, Simic M, Pariante CM, Papadopoulos A, Cleare A, Brammer M, Fombonne E, Rubia K (2009) Reduced activation in lateral prefrontal cortex and anterior cingulate during attention and cognitive control functions in medication-naive adolescents with depression compared to controls. J Child Psychol Psychiatry 50:307–316

Henter ID, de Sousa RT, Zarate CA Jr (2018) Glutamatergic modulators in depression. Harv Rev Psychiatry 26(6):307–319

Heresco-Levy U, Gelfin G, Bloch B, Levin R, Edelman S, Javitt DC, Kremer I (2013) A randomized add-on trial of high-dose D-cycloserine for treatment-resistant depression. Int J Neuropsychopharmacol 16:501–506

Hong Ng C, Norman TR, Naing KO, Schweitzer I, Kong Wai Ho B, Fan A, Klimidis S (2006) A comparative study of sertraline dosages, plasma concentrations, efficacy and adverse reactions in Chinese versus Caucasian patients. Int Clin Psychopharmacol 21:87–92

Howland RH (2008) Sequenced Treatment Alternatives to Relieve Depression (STAR*D). Part 2: Study outcomes. J Psychosoc Nurs Ment Health Serv 46:21–24

Ide S, Ikeda K (2018) Mechanisms of the antidepressant effects of ketamine enantiomers and their metabolites. Biol Psychiatry 84:551–552

Iosifescu DV (2015) Maintaining the initial clinical response after ketamine in bipolar and unipolar depression: an important next-step challenge. J Clin Psychiatry 76:738–740

Iosifescu DV et al (2009) Frontal EEG predictors of treatment outcome in major depressive disorder. Eur Neuropsychopharmacol 19:772–777

Jesulola E, Sharpley CF, Bitsika V, Agnew LL, Wilson P (2015) Frontal alpha asymmetry as a pathway to behavioural withdrawal in depression: research findings and issues. Behav Brain Res 292:56–67

Khodayari-Rostamabad A, Reilly JP, Hasey GM et al (2013) A machine learning approach using EEG data to predict response to SSRI treatment for major depressive disorder. Clin Neurophysiol 124:1975–1985

Lally N, Nugent AC, Luckenbaugh DA, Niciu MJ, Roiser JP, Zarate CA Jr (2015) Neural correlates of change in major depressive disorder anhedonia following open-label ketamine. J Psychopharmacol 29:596–607

Leisman G, Moustafa AA, Shafir T (2016) Thinking, walking, talking: integratory motor and cognitive brain function. Front Public Health 4:94

Li CT, Su TP, Wang SJ, Tu PC, Hsieh JC (2015) Prefrontal glucose metabolism in medication-resistant major depression. Br J Psychiatry 206:316–323

Li CT, Chen MH, Lin WC, Hong CJ, Yang BH, Liu RS, Tu PC, Su TP (2016) The effects of low-dose ketamine on the prefrontal cortex and amygdala in treatment-resistant depression: a randomized controlled study. Hum Brain Mapp 37:1080–1090

Lin KM, Shen WW (1991) Pharmacotherapy for southeast Asian psychiatric patients. J Nerv Ment Dis 179:346–350

Lin CT et al (2008) Noninvasive neural prostheses using mobile and wireless EEG. Proc IEEE 96:1167–1183

Melo A, Leite-Almeida H, Ferreira C, Sousa N, Pego JM (2016) Exposure to ketamine anesthesia affects rat impulsive behavior. Front Behav Neurosci 10:226

Murrough JW, Iosifescu DV, Chang LC, Al Jurdi RK, Green CE, Perez AM, Iqbal S, Pillemer S, Foulkes A, Shah A, Charney DS, Mathew SJ (2013) Antidepressant efficacy of ketamine in treatment-resistant major depression: a two-site randomized controlled trial. Am J Psychiatry 170:1134–1142

Murrough JW, Soleimani L, DeWilde KE, Collins KA, Lapidus KA, Iacoviello BM, Lener M, Kautz M, Kim J, Stern JB, Price RB, Perez AM, Brallier JW, Rodriguez GJ, Goodman WK, Iosifescu DV, Charney DS (2015) Ketamine for rapid reduction of suicidal ideation: a randomized controlled trial. Psychol Med 45:3571–3580

Nachev P, Kennard C, Husain M (2008) Functional role of the supplementary and pre-supplementary motor areas. Nat Rev Neurosci 9:856–869

Petryshen TL, Sabeti PC, Aldinger KA, Fry B, Fan JB, Schaffner SF, Waggoner SG, Tahl AR, Sklar P (2010) Population genetic study of the brain-derived neurotrophic factor (BDNF) gene. Mol Psychiatry 15:810–815

Phillips ML, Chase HW, Sheline YI, Etkin A, Almeida JR, Deckersbach T, Trivedi MH (2015) Identifying predictors, moderators, and mediators of antidepressant response in major depressive disorder: neuroimaging approaches. Am J Psychiatry 172:124–138

Ressler KJ, Mayberg HS (2007) Targeting abnormal neural circuits in mood and anxiety disorders: from the laboratory to the clinic. Nat Neurosci 10:1116–1124

Singh JB, Fedgchin M, Daly EJ, De Boer P, Cooper K, Lim P, Pinter C, Murrough JW, Sanacora G, Shelton RC, Kurian B, Winokur A, Fava M, Manji H, Drevets WC, Van Nueten L (2016) A double-blind, randomized, placebo-controlled, dose-frequency study of intravenous ketamine in patients with treatment-resistant depression. Am J Psychiatry 173:816–826

Su TP, Chen MH, Li CT, Lin WC, Hong CJ, Gueorguieva R, Tu PC, Bai YM, Cheng CM, Krystal JH (2017a) Dose-related effects of adjunctive ketamine in taiwanese patients with treatment-resistant depression. Neuropsychopharmacology 42:2482–2492

Su TP, Chen MH, Chang WH, Li CT, Lin WC, Hong CJ, Tu PC, Bai YM, Cheng CM (2017b) Dose-related effects of ketamine infusion on suicidal ideation in patients with treatment-resistant depression: a randomized double-blind controlled study in Taiwan, 5th AsCNP biennial meeting, Bali, Indonesia

Wan LB, Levitch CF, Perez AM, Brallier JW, Iosifescu DV, Chang LC, Foulkes A, Mathew SJ, Charney DS, Murrough JW (2015) Ketamine safety and tolerability in clinical trials for treatment-resistant depression. J Clin Psychiatry 76:247–252

Wang J, Zhou M, Wang X, Yang X, Wang M, Zhang C, Zhou S, Tang N (2014) Impact of ketamine on learning and memory function, neuronal apoptosis and its potential association with miR-214 and PTEN in adolescent rats. PLoS One 9:e99855

Wang LJ, Chen CK, Lin SK, Chen YC, Xu K, Huang MC (2018) Cognitive profile of ketamine-dependent patients compared with methamphetamine-dependent patients and healthy controls. Psychopharmacology (Berl) 235:2113–2121

WHO (2008) Projections of mortality and burden of disease, 2004–2030

Wilkinson ST, Ballard ED, Bloch MH, Mathew SJ, Murrough JW, Feder A, Sos P, Wang G, Zarate CA Jr, Sanacora G (2018) The effect of a single dose of intravenous ketamine on suicidal ideation: a systematic review and individual participant data meta-analysis. Am J Psychiatry 175:150–158

Zanos P, Gould TD (2018) Mechanisms of ketamine action as an antidepressant. Mol Psychiatry 23:801–811

Zarate CA Jr, Brutsche N, Laje G, Luckenbaugh DA, Venkata SL, Ramamoorthy A, Moaddel R, Wainer IW (2012) Relationship of ketamine's plasma metabolites with response, diagnosis, and side effects in major depression. Biol Psychiatry 72:331–338

The manufacturer's authorised representative in the EU is Springer Nature Customer Service Centre GmbH, Europaplatz 3, 69115 Heidelberg, Germany. If you have any concerns regarding our products, please contact ProductSafety@springernature.com

Printed and bound by CPI Group (UK) Ltd, Croydon, CR0 4YY

30/09/2026

02225449-0001